THE CRIMES OF DIGITAL CAPITALISM

JUSTICE, INEQUALITY, AND THE DIGITAL WORLD
General Editors: Jan Haldipur and Calvin John Smiley

The Crimes of Digital Capitalism: Corporate Crime in an Age of Exploitation
Aitor Jiménez

The Crimes of Digital Capitalism

Corporate Crime in an Age of Exploitation

Aitor Jiménez

NEW YORK UNIVERSITY PRESS
New York

NEW YORK UNIVERSITY PRESS
New York
www.nyupress.org

© 2025 by New York University
All rights reserved
Library of Congress Cataloging-in-Publication Data
Names: Jiménez, Aitor, 1986– author.
Title: The crimes of digital capitalism : corporate crime in an age of exploitation / Aitor Jiménez.
Description: New York : New York University Press, [2025] | Includes bibliographical references and index.
Identifiers: LCCN 2024020143 (print) | LCCN 2024020144 (ebook) | ISBN 9781479821693 (hardback) | ISBN 9781479821716 (paperback) | ISBN 9781479821761 (ebook) | ISBN 9781479821723 (ebook other)
Subjects: LCSH: Big business—Moral and ethical aspects. | Capitalism—Moral and ethical aspects. | Corporations—Moral and ethical aspects. | Information technology—Moral and ethical aspects.
Classification: LCC HD2351 .M37 2025 (print) | LCC HD2351 (ebook) | DDC 338.6/44—dc23/eng/20240815
LC record available at https://lccn.loc.gov/2024020143
LC ebook record available at https://lccn.loc.gov/2024020144

New York University Press books are printed on acid-free paper, and their binding materials are chosen for strength and durability. We strive to use environmentally responsible suppliers and materials to the greatest extent possible in publishing our books.

Manufactured in the United States of America

10 9 8 7 6 5 4 3 2 1

Also available as an ebook

Para Gala y Nuño, porque ellos crean mundo

CONTENTS

Introduction: Automating the Crimes of the Powerful — 1

1. The Digitization of State Racism — 31
2. The Digital Takeover of Education — 54
3. Cyberwar against the People — 85
4. The Materiality of Digital Exploitation — 115
5. Law and Extractivism — 133
6. Killing the Salar de Atacama — 163

Conclusion: Platformed Criminals — 185

Acknowledgments — 207

Notes — 209

Bibliography — 227

Index — 261

About the Author — 277

Introduction

Automating the Crimes of the Powerful

"Amazon won't let us go" were the last words Larry Virden was able to write to his girlfriend on December 10, 2021. A few minutes later, the roof of the Amazon fulfillment center in Edwardsville, Illinois, where Virden worked, collapsed, killing him and five other coworkers and injuring one other worker.[1] This was no sudden tragedy, no unforeseen catastrophe. Thirty-six hours earlier, the weather services had already warned the company of the dangerous weather forecasts: storms and tornadoes. Despite this, Amazon did not restructure its workers' schedules or delivery routes. This was a peak period before Christmas, with a huge volume of business that demanded extra shifts from its precarious workers. On December 10, the company received numerous messages and calls from its delivery drivers informing them of the dangerous conditions. Just an hour and a half before the warehouse collapsed, Amazon was still demanding its workers to "keep delivering" or else face layoffs. The following is an excerpt from a conversation between an Amazon delivery worker in Illinois and his line manager:

> DRIVER: Tornado alarms are going off over here.
> DISPATCH: Just keep delivering for now. We have to wait for word from Amazon. If we need to bring people back, the decision will ultimately be up to them. I will let you know if the situation changes at all. I'm talking with them now about it.
> DRIVER: How about for my own personal safety, I'm going to head back. Having alarms going off next to me and nothing but locked building [sic] around me isn't sheltering in place. That's wanting to turn this van into a casket. Hour left of delivery time. And if you look at the radar, the worst of the storm is going to be right on top of me in 30 minutes.

DRIVER: It was actual sirens.
DISPATCH: If you decided to come back, that choice is yours. But I can tell you it won't be viewed as for your own safety. The safest practice is to stay exactly where you are. If you decide to return with your packages, it will be viewed as you refusing your route, which will ultimately end with you not having a job come tomorrow morning. The sirens are just a warning.
DRIVER: I'm literally stuck in this damn van without a safe place to go with a tornado on the ground.[2]

Amazon did not implement any emergency plan, did not evacuate its workers. Profit first, safety second. Between 8:06 p.m. and 8:16 p.m. the company received calls warning of the impending tornado impact. Amazon reacted by advising its workers to take shelter in its warehouse bathrooms. At around 8:20 p.m. the tornado ripped through the facility. Despite the dead and injured, the lawsuits and the protests of the families, Amazon went unpunished; for the state, it was just a tragic accident. It was not the first time that workers of digital platforms were forced to perform their tasks in a situation of uncertainty and serious risk.

It may seem remote to you now, but think back for a moment to the first half of 2020. The era of global shock. That was the year of the pandemic, the year when almost everything stopped. It was also the time when digital and analog life became one (or rather, when we realized it). I remember the anxiety with which I followed the news. Little or nothing was known about a disease that seemed to have no limits. I remember my mother on the phone, telling me how the ambulance sirens in Madrid did not stop, sounding day and night. I remember when the government declared a state of emergency and lockdowns. I remember reading nervously about the growing number of infections and deaths. I remember the uneasiness I felt just going out to do the most basic tasks. I remember the security measures, the police roadblocks, the drones and helicopters flying over the cities. I remember how quickly concepts like distance learning and telecommuting became widespread and how my social life was suddenly reduced to the size of my phone. Among the many scenes of that disaster, there is one thing that particularly stuck with me, and that is an image.

This image shows emergency medical and military personnel setting up a health checkpoint somewhere in the city. Following the regulated protocols indicated by the government, the officers were dressed in biosafety suits. It was not a manga comic book or a video game. It was not a science fiction movie. It was reality, live and direct. But there was something that didn't fit that image. In contrast to the Hollywood scene, another figure could be seen. A bicycle messenger. On one side of the image is a more or less adequate representation of the catastrophe. On the other side is a worker "protected" by a mask and blue plastic gloves, transporting food from some restaurant to some customer in the city. The scene took place in Barcelona, but it equally could have happened in New York, Los Angeles, Toronto, Buenos Aires, or Paris.

COVID-19 is now part of the long list of viruses we live with, but at the beginning of 2020 things were very different. Hardly anything was known about the transmission, extent, or potential short-, medium- and long-term repercussions of the disease. The idea of a vaccine was not on the near horizon. Attention was fixed on the numbers of infections, the doctors available, and the insufficient number of ventilators for the thousands of people whose lives were at stake. There was fear and there was anxiety. It was not just about personal safety, people feared that a gesture as natural as a hug or a kiss could put people at risk. In places as far away as Rio, New York, Madrid, or Melbourne, a whole generation of gig workers became essential workers overnight. But they were not workers like everyone else. The new labor regime imposed by tech companies like Deliveroo, Uber, and Amazon has placed gig workers outside the old social pact of work[3]—that is, outside the regime of social benefits, access to unemployment, and health benefits that, despite the neoliberal tempest, other workers still enjoy. The gig work regime is one of insecurity, digital surveillance, and wages (often piece rate) below minimum income, in which associating, unionizing, and striking are possibilities proscribed by the very technological architecture of the platforms. It is something, nevertheless, that is also the last resort for hundreds of thousands of impoverished people, the vast majority of whom are migrant and racialized.[4] They are an unprotected workforce submerged in the continuing global crisis and the worsening of the material conditions of existence. In the words of one gig worker, "All gig workers who are working right now are working because they can't afford not to: they live week to week and there is no backup."[5]

Yes, it is true, some of the measures that were taken have proven to be overzealous and restrictive. Perhaps we would do things differently now. But the point is not to analyze retrospectively with today's knowledge of the actual degree of exposure to a biological threat that the gig workers had. What is important here is to see how, in a context of general alarm, uncertainty and risk, digital capitalist companies showed absolute disdain for the safety of workers who managed to keep their jobs. Meanwhile, those who saw their shifts drastically reduced or their employment relationship terminated were abandoned to their fate since, in many cases, the company did not even consider them as workers. I am sure that in the event of a hypothetical alien invasion and subsequent societal collapse, we would still be able to receive—in the comfort of catastrophe—sushi, pizza, or tikka masala delivered by couriers.

For me this is not an anecdote. It is proof of how digital capitalism increases, multiplies, and creates devastating forms of social, economic, political, cultural, and environmental harm under a hype discourse of creativity, innovation, and supposed sustainability. This book offers an exploration of forms of state and corporate criminality in the digital age through the analysis of three key but converging axes: the rise of corporate power and the regime of impunity and state complicity in which it operates; the infrastructure of global racial capitalism that structures relations of global productive domination; and the spectacular rise, ubiquity, and societal dependence on digital technologies. These are the fundamental ingredients of what I call in this book the crimes of digital capitalism.

Those familiar with the literature on state and corporate crimes (something I further identify as the crimes of the powerful) may be asking, well, what's new? After all, there is already a wealth of literature demonstrating how global capitalism is built on the racial hierarchization of labor and environmental destruction and how large corporations profit from the privatization of public services, imperialist wars, the construction of prisons—in addition to literature discussing the establishment of a global regime of surveillance, control, and punishment in general. Without abandoning the study of these logics that are unfortunately still active, this book aims to resituate the discussions of critical criminology that analyze state and corporate crimes in the face of the reality of the digital era. The aim of the following chapters is to expose how digital capitalism turbo-accelerates, multiplies, extends, and invisibilizes—under the cloak of sustainability and philanthropy—old and new forms of social harm caused by states and corporations while also creating new scenarios and mechanisms for privatization, exploitation, and domination. If there is one field in which digital capitalism shows both its continuity with other forms of domination and its specificity, it is in the intensity of and dependence on digital technologies.

The Platform Society

I guess you've noticed it. If you're over thirty, you know there was a time, a space where you would wake up without reaching out for your cell phone. A time when you could usually meet someone at a certain time somewhere and be (sometimes) certain that they would be there without the need for constant, real-time confirmation. It seems difficult today to imagine a world without Google Maps—how did people find the streets, the restaurants, the way home, to a mountain, to the corner store? A world without streaming. A world without the need to communicate your latest unfiltered thought on a social network owned by a racist megalomaniac billionaire. A world where it was not necessary to have an app to confirm that *you* were *you* to access financial services, or to search the university library, or to fill out taxes. A world without 24–7 availability. A world where vacation planning did not go through the algorithmic filter of a company capable of offering the largest hotel offer in the world (and profiting from it) without having a single establishment for lodging. A world where most new relationships had their first point of contact in other friends and not a proprietary algorithm. Digital capitalism has fluidized communications, commerce, sex, work. For many, digitization has allowed the acceleration of processes and access to resources and people previously distant. In a way, it is the realization of the Italian futurists' old-new dream: "We declare that the splendour of the world has been enriched by a new beauty: the beauty of speed. A racing automobile with its bonnet adorned with great tubes like serpents with explosive breath . . . a roaring motor car which seems to run on machine-gun fire is more beautiful than the Victory of Samothrace."[6]

Today, speed is measured by the capacity to transmit information in 5G, 6G, or 7G, but the veneration for technique and the spirit of time is the same. The elimination of frictions, ties, roots; historical nemeses of capitalism and fascism. "Move fast and break things," they said at Facebook.[7] This destruction is announced as emancipating, yet it establishes new verticalities and centralizations, of information, of power, of privilege. It is a destruction that is to a certain extent necessary—can anyone defend the morality or ethics of the old corporations and states of the twentieth century?—but nevertheless also devastates everything that is autonomous and communitarian while multiplying a regime of extensive, global and granular power. I do not know if it was a better world, but it is undeniable that it was different. And perhaps the most relevant fact here is not so much the undeniable expansion of the technological mediation of our societies as it is the power acquired by the agents that design, manage, and monetize it. Yes, I am talking about the big technology companies; I am talking about digital capitalism and its presence in our lives.

There is no doubt that we find ourselves immersed in a social fabric dependent on the services provided by a small group of technological corporations. This phenomenon, which encompasses both the individual and collective spheres, is evidence of the consolidation of a power that goes beyond the market and the supply of services to situate itself in the social and productive fabric of our era. It is a power with the capacity to substantially influence the configuration of our individual and collective lives. And no, it is not a matter of hocus pocus, mind hacking, or esoteric algorithms. It is a case of something as material as corporate control over the infrastructures of the everyday. The examples are numerous. According to the United States Congressional Research Service, 99 percent of interoceanic Internet traffic takes place on cables owned by large corporations.[8] This proprietary power is exercised by corporations such as Amazon, Alphabet, China Mobile, and Meta and extends to data centers, information capture systems, the computational power to process it and, of course, the technological and human capacity to produce and develop new technologies.

This infrastructural power, obtained through legal and illegal practices such as abuses against competition, constant violations of privacy, labor exploitation, and absolute disregard for the environment, is behind the hegemonic power enjoyed by the major technology companies. A clear indicator of this hegemony can be seen in the field of information and communication. In this scenario, companies such as Alphabet and Meta not only stand as undeniable leaders but practically become the guardians of digital truth. More than 90 percent of searches in the United States are conducted through Google, and about 30 percent of US adults rely on Facebook as their primary source of information.[9] The Android operating system (which prioritizes the use of Google over other search engines) is present on 97 percemt of all smartphones in India. That is, just over half a billion people in that country alone spend part of their digital lives in a proprietary digital enclosure. These numbers reveal an extensive and profound influence not only on the way we inform ourselves, but also indicate corporate control over the architecture of access and construction of knowledge.

When we enter the entertainment arena, it is impossible to ignore the cultural dominance exerted by platforms such as Netflix, Spotify, Disney, Amazon Prime, HBO, and Apple. These companies have transcended the simple provision of streaming services, becoming architects of global tastes and trends. They have become the cultural curators of half of humanity. Their impact on shaping popular culture is reflected not only in the number of users, but also in how these platforms shape conversations, ideas, tastes, and aesthetics.[10] This process is not limited to communication and access to culture and information.

In a span of just fifteen years, Amazon has emerged as one of the main arbiters in the movement of goods in the Global North. The notion of a free market pales in comparison to the dominance this company exerts over online sales, logistics, and, of course, pricing. Market dominance not only impacts competition but redefines the role assumed by users and workers. The shopping experience for millions of individuals—from product choice to delivery—is completely mediated by an opaque algorithmic architecture that spies on users while offering them dynamic and geolocated prices—a situation of information asymmetry that puts customers in a position of disadvantage and helplessness. In terms of work, Amazon has profoundly transformed the production chain of the retail sector. The transformation begins in the distribution and operation of its logistics centers where thousands of workers, under a regime of algorithmic and robotic surveillance, work in support of the warehousemegamachine. The labor transformation continues with the distribution of goods, not by workers but by freelancers, supposedly "independent" employees, who will make their deliveries following the instructions of a precarious assembly of guidance and work management devices. Wiped off the map are numerous retail stores that find it impossible to compete with the market penetration of the digital giant. However, behind the facade of logistical efficiency lies a more complex reality. The carbon emissions derived from its extensive logistics network, as well as the waste management generated by its packaging, generate a profound ecological impact. Amazon's own data reveal that its carbon emissions were equivalent to approximately seventy-two million metric tons in 2022. Its business model, despite the narrative of decarbonization, is totally dependent on fossil fuels.[11]

In parallel, labor disputes have marked Amazon's trajectory. Strikes and worker protests have broken out in different parts of the world, evidencing significant tensions related to ominous working conditions characterized by intensive surveillance, antiunion policies, and wage theft. For example, in 2020, during the COVID-19 pandemic, Amazon warehouse workers organized protests in several countries, demanding adequate safety measures and better working conditions.[12] More recently, thousands of workers at Amazon centers in Europe went on strike on Black Friday 2023 under the slogan "Make Amazon Pay." In addition, Amazon's conflicts with various governments have become a constant. From tax disputes to regulatory tensions, Amazon's global expansion has generated debates and confrontations with government authorities. The materiality of domination is further evidenced by the clarity of corporate control over the digital world in which we spend a good part of our lives. The "personalization" and prioritization of content in our social networks, recommendations in search engines, and hypertargeted (or hidden) advertising, are examples of how big technology subtly shapes our perception of reality, creating an asymmetrical relationship between individuals and technological giants in a world where everyday decisions are increasingly mediated by proprietary algorithms.

But corporations do not work alone. As writer Antony Loewenstein has recently pointed out, myriad Palestinian and Arab content producers have encountered human and algorithmic censorship of images, messages, and videos at the hands of corporations like Facebook.[13] Numerous accounts reporting on the reality of the occupied territories in the West Bank or Israeli state terrorism in Gaza were deleted and canceled from platforms such as Youtube, Instagram, and Facebook without explanation or on the grounds of collusion with terrorist causes as directed by the Israeli state. The technological silencing of critical voices coming from cultural and religious minorities and their singling out by oppressive majorities (as is the case of India and the Muslim population of Kashmir, and of the Rohingya in Myanmar) are not exceptions but an everyday occurrence, something that contrasts with the permissiveness enjoyed by oppressors. While the cancellation of pro-Palestinian content took place, the reproduction and viralization of content celebrating bombings and assassinations perpetrated against the Palestinian people and disseminated by the State of Israel itself was allowed.[14]

Behind this selective censorship lie not technical errors but policies agreed between corporations like Meta and states like Israel whose bargaining power with large corporations (guided exclusively by the interest of market access) puts them in an advantageous position to influence and modulate dissenting voices in the supposedly independent and free digital spheres. These relationships were apparent during the Israeli invasion of Gaza in late 2023 where in the midst of the Palestinian genocide (fifteen thousand dead in the first two months of the war alone) Elon Musk owner of X (formerly Twitter) held a meeting with the Israeli prime minister and architect of the invasion, Benjamin Netanyahu, who received the billionaire with the publicity and protocols befitting a head of state. From the meeting it was learned that Elon Musk pledged not to provide Internet to Gaza through its Starlink satellites without Israel's approval.[15] In other words, a corporation bowed to the censorship wishes of an occupying power that has been accused of crimes of genocide. These examples underscore how technological corporations' control over digital infrastructures gives them enormous and pernicious power over our individual and collective lives—a power that extends from the shaping of our physical and digital experiences to the control and censorship of resistance practices or state policies. This obviously has significant impacts, ranging from political polarization, censorship, and control of markets and workers to discrimination, racism, and cooperation with old and new forms of colonialism and oppression. But, as I have already begun to point out, corporations could not enjoy this power without an appropriate legal and political architecture—that is, without the complicity of states.

Techno-Solutionism and Politics

The year 2023 was the warmest year on record, surpassing the recent records of 2015 and 2016.[16] It is difficult to prove, but some data suggest that the world has not experienced such high temperatures in the last 125,000 years, and the outlook is only more and faster warming. From a geological point of view this is neither a climatic extravagance nor a surprise. After all, we are living in the Holocene, a warm geological era that succeeded the last ice age that occurred about 10,000 years ago. The problem is not warming per se. Earth has gone through warm and cold periods; the climate is subject to variations and transitions. The problem is the speed, the acceleration of change. This has resulted in unprecedented droughts in the United Kingdom in 2023 and devastating floods like those in Pakistan in 2022. The world is experiencing a fracture. Recognizing this is not catastrophism, it is *realpolitik*. In the words of United Nations (UN) secretary general António Guterres, "We are living through climate collapse in real time—and the impact is devastating."[17] There is a scientific consensus that the proliferation of greenhouse gases and the reduction of forest cover are physical causes of the rapidity of these changes, both phenomena linked to human activity. In other words, global warming is a natural process that has been turbo-accelerated since the industrial era, especially in the last hundred years. A phenomenon intimately linked to what activist and researcher Andreas Malm calls *fossil capital*.[18] Fossil capital is a regime of capitalist production and exploitation dependent on extractivism, particularly the consumption of fossil fuels. It is a hypercentered capitalism with enormous power. According to some estimates by the Climate Accountability Institute,[19] just one hundred companies are responsible for 71 percent of CO_2 emissions since 1988. Twenty-five of the largest are responsible for almost 50 percent of emissions.[20] This is not an innocent regime that has operated in ignorance. The big beneficiaries of fossil capitalism—states, energy, and automobile companies, such as Texaco, Ford, GM, and ExxonMobil—have known for at least fifty years about the potentially devastating consequences of "business as usual" for the environment and for human societies.[21] These consequences were something they tried to deny, then obfuscate, and then finally consider as a fait accompli and rebrand as necessary for development, progress, and civilization.

In other words, we know that we are facing a socio-environmental disaster with a clear point of origin. This is something that is recognized not only in academic or activist circles. Political institutions such as the European Union (EU) and the UN have recognized and pointed in countless studies to the origins and responsibilities of the catastrophe caused by fossil capitalism. These same institutions recognize the alarming climate projections and the impact it will have on regions such as the Mediterranean, especially fragile from an environmental and social point of view. It is not a future problem. In 2016, there was already talk of the worst drought in the western Mediterranean in nine hundred years.[22] In 2022, countries such as Morocco, Algeria, Spain, France, and Italy experienced temperatures between 2.5 and 4 degrees Celsius above average. Morocco, Algeria, Tunisia, and large regions of Spain have been under drought alert throughout 2023.[23] The city of Barcelona is already making calculations to provide water to the city by tankers. What solutions are on the table right now? What are the major projects and measures to combat this structural threat? Perhaps to stop capitalist logics based on the doctrine of economic development and eternal growth? To stop the extractivist policies that have led in the first instance to the current emergency? To listen to the critical voices that for decades have been denouncing and confronting the expansion of infrastructures of destruction, such as pipelines, extraction sites, and processing plants? To respect the Indigenous peoples and peasants who for centuries have been resisting incorporation into capitalist modernity and its ecocidal logics? No. Why renounce capitalism if capitalism itself offers the solution? Perhaps fossil capitalism is leading to a dead end—but there is an alternative, European and American institutions seem to say: green and digital capitalism.

Leaders from across the political spectrum believe they have found in digital technologies the solution to the structural, environmental, social, and economic problems that are shaking the world. For example, the EU has mobilized the largest amount of resources in its history to launch the so-called twin transitions to the "green and digital." The Global North has surrendered its political hopes to technological solutionism, that is, the unfounded belief that "technology" will solve societal challenges without the need for major structural changes. A supposed panacea policy that will simultaneously reindustrialize and decarbonize the depressed economies of Europe and the United States without questioning the dogmas of capitalism: growth, production, and consumption. Apparently we can save the world by replacing combustion-engine cars with Tesla sports cars and Volkswagen electric SUVs and by teleworking from our microapartments in gentrified neighborhoods while connected to cell phones and terminals via 6G towers. Surely most Europeans will even be able to enjoy the six summer months of the new climate era on the beaches of Portugal and Spain with the clear conscience of having used biofuel-powered airplanes or been transported in high-speed trains. Moreover, the rural territories suffering from the severe drought in a large part of the Mediterranean will not have to question the agro-export model based on semiarid land irrigation and fossil-fuel-addicted distribution. It will be enough to automate and electrify agricultural machinery and to plague crops and irrigated land with sensors and drones. As I discuss below, the green and digital transition does not question the extractive logics that have led to the current emergency. As recognized by its main protagonists, it is virtually impossible to carry out this transition without exponentially increasing the extraction and processing of natural resources.[24] This is confirmed by the new wave of legislation on materials, industries, and critical infrastructures that seeks to speed up the extraction and refinement of raw materials such as lithium. It is no wonder that the major players in the energy and mining industry are the main advocates of green and digital transitions.[25]

However, the main problem lies not in the big headlines that promise huge transformations with little friction, change, or suffering but in the dense processes that are taking place unnoticed. Public institutions are increasingly dependent on the critical infrastructures of the digital age. Proof of this is the growing number of administrations that entrust the management and storage of the data of hundreds of millions of citizens to companies such as Amazon and Microsoft. It is a textbook example of the privatization of public administration.[26] Similarly, Alphabet has been the protagonist and main beneficiary of what may be the fastest and most successful privatization of a public service in history. Alphabet now controls the data of more than 150 million students, not to mention the access this indirectly allows into their families and teachers, and this is just one of many educational platforms. As Niels Kerssens and Jose Van Dijck demonstrate, educational infrastructures are being captured by a handful of digital corporations.[27] Corporate dominance of educational technologies goes beyond control over the means of information transmission. Platformized education is reshaping the way we understand pedagogy, its meaning, purpose, and goals, now reinterpreted through neoliberal technologies. The takeover of key state functions—health, education, and security—by a cartel of corporations represents a significant threat to our ability to protect and care for our communities. Aptly, Benvegnù and colleagues ask, "Is it right that an infrastructure so fundamental to social reproduction be managed by a subject that is neither accountable nor guided by the logics of the common good?"[28]

Within the intricate web of algorithms that exert control over our public and private lives, automated decision-making systems occupy a central place. *Automated decision-making system* refers to any software, system, or process designed to automate, assist, or replace human decision-making.[29] Ongoing research conducted by MuckRock and the Rutgers Institute for Information Policy and Law reveals that US government bodies employ approximately 203 automated decision-making systems at various levels.[30] Another study commissioned by the Administrative Conference of the United States, conducted by Stanford and New York Universities, revealed that nearly half (45 percent) of the 142 agencies surveyed used some type of automated decision-making system. These new government technologies span a wide range of policy areas, including law enforcement, healthcare, financial regulation, social welfare, commerce, environment, science and energy, communications, agriculture, labor and employment, transportation, housing, and education.[31]

The proliferation of algorithmic tools in government systems is exacerbating old discriminatory practices while inflicting new social harms. As early as 2019, the UN special rapporteur on extreme poverty and human rights warned that "as humanity moves, perhaps inexorably, into the future of digital well-being, it needs to alter course significantly and quickly to avoid stumbling zombie-like into a digital well-being dystopia."[32] Such a scenario does not stem from technological errors or faulty data sets, as demonstrated by the implementation and consequences of algorithmic justice and predictive policing in New York City.³³ On the contrary, as I will illustrate in chapters 2 and 4, this regime of "punitive capitalism" is the predictable outcome of socio-technical systems tailored to regimes that rely on indiscriminate mass surveillance, the repression of refugees, and the policing of poor and racialized communities as a form of governance.³⁴

History repeats itself; capitalism has used an episode of crisis to implement a radical transformation. The COVID-19 pandemic functioned as an extraordinary transmitter of the Silicon Valley shock doctrine.³⁵ The Screen New Deal, negotiated worldwide between tech billionaires and governments in crisis, demonstrated that the distinction between the digital and the "real" has completely disappeared.³⁶ To govern in one place is to govern in the other. The close relationship between digital capitalism and State apparatus points to a qualitative leap in governmental strategies that are rapidly moving towards new forms of algorithmic governance. Despite being discredited during the massive corporate theft that was the 2008 crisis, neoliberalism has found a way to hack the socio-technical systems of the postpandemic era. We are witnessing the solidification of governmental techniques rooted in cybernetics and neoliberalism, both foundational layers of digital capitalism. Western democracies, broadly defined by the canonical elements of liberal ideology—democratic elections, representative government, separation of powers, state sovereignty, individual and collective rights, and rule of law—are being hacked. A new constellation of epistemologies, practices, and beliefs (such as digital democracy and algorithmic regulation) is devouring from within what remains of the welfare states of the Global North. As investor and tech guru Marc Andreessen prophetically announced several years ago, "Software is eating the world."³⁷

Numerous discussions have already shed light on the corporations spearheading this transformative change as well as on how digital capitalism intensifies the exploitation of labor,³⁸ commodifies rural and urban spaces, privatizes public domains, poses a threat to democracy,³⁹ surveils political dissenters,⁴⁰ and employs greenwashing tactics to cover up these processes.⁴¹ Nick Dyer-Whiteford calls attention to the "inhuman power" wielded by artificial intelligence (AI) capitalism,⁴² noting how cyberlibertarian assembly is capturing the critical infrastructures of tomorrow. Meanwhile, Dan McQuillan called for anti-fascist unity to resist against the necropolitics and violences linked to the political economy of AI. More concretely, Felicity Amaya Schaeffer and Petra Molnar have demonstrated the convergence between technologization and racialized militarization of borders.⁴³

Mark Andrejevic was one of the first to approach the convergence between surveillance, technology, and capitalism, a phenomenon he called digital enclosure, or a form of value capture and monetization of human experience aimed at accumulation for the benefit of a nascent digital ownership class.⁴⁴ Others emphasize how digital capitalism exercises an almost unquestionable cultural hegemony in society, creating a situation of digital alienation that prevents us from being able to imagine a postcapitalist future.⁴⁵ For their part, Nick Couldry and Ulises Mejias show how, thanks to the widespread presence of devices and the unprecedented computational power held by large corporations, "data are colonizing human life and appropriating it for capitalism."⁴⁶ For its part, the Tiqqun collective already unveiled a few years ago how big tech captured and actualized the economy of desire of late capitalism, hijacking our potential for the common production of knowledge, while we submit ourselves to the voluntary servitude of technologies that are literally killing us.⁴⁷

Interest in the considerable social harm caused by the operations of digital capitalism has transcended the confines of academia to reach the public consciousness. Rare is the day when there is not an article highlighting scandals related to the big tech companies.⁴⁸ As an example, in December 2023, the Asociación de Medios de Información de España, representing the Spanish newspaper employers' association, filed a lawsuit against Meta for unfair competition, replicating similar moves that took place in Australia and Canada. Even governments previously favorable to the big tech companies are now taking this issue seriously and trying to control an industry that has firmly consolidated its leading position in all sectors, countries, and territories.⁴⁹ Examples abound. In the United States alone, forty-one states had initiated legal proceedings against Facebook-Meta by 2023, alleging among other things that the company routinely captures data from children under the age of thirteen without parental consent.⁵⁰

Digital capitalists have deliberately lied to representatives around the world. They have shown deep disdain for the (limited) wealth redistribution tools in the countries where they operate, devising sophisticated forms of tax evasion or simply illegally evading taxes.[51] They have massively collected and exploited data to then illegally trade with it, thereby building the base on which their quasi-monopolistic power rises. They have abused their dominant position, destroying and buying out potential competitors. In short, they have consciously caused extensive social damage by putting profits before the common good.

Numerous questions arise. What motivates authorities and elected representatives around the world to compromise, negotiate, and place our security, data, health, and education in the hands of corporate entities with pronounced psychopathic tendencies? Why, despite high-level political statements, hearings, and parliamentary inquiries, has so little been done in effective terms to address corporate power? Why, despite incredible and ever-increasing research uncovering the infinite ways in which digital corporations cause social harm in every conceivable way, do we still lack a comprehensive criminal theory fit for the digital age? Why do the socially harmful behaviors of "digital power brokers" so often go unpunished and, more to the point, become naturalized as necessary consequences of bringing about the "green and digital transition"? For me, these are not abstract issues arising from an exclusively academic interest. These questions have guided and fueled my political interest over the last eight years, during which I have collaborated with social organizations, political parties, and unions trying to confront the umpteenth capitalist reconfiguration. Answering these questions will be the aim of this book; to do so, I will make use of the theoretical and methodological tools provided by criminology, sociology, and critical legal studies.

The Crimes of the Powerful

On September 23, 2021, I had the enormous privilege of testifying as an expert in the national security commission of the Spanish congress. The topic to be discussed was the possible social risks derived from online behaviors, especially the spreading of disinformation and fake news. The focus of the debate was placed within the parameters of security. The military spoke, mentioning the need to strengthen cyber defenses against foreign interference. Experts testified, sharing their knowledge on the rise of radicalism and terrorism on the Internet. They spoke of the contamination of the online public space, the spread of hatred, and the vulnerabilities of open societies in a hyperconnected world. They discussed the role and dangers of news verification agencies. But despite the change of register, the discourse followed the traditional line followed until then by much of the discussions and policies in different European countries and whose argument could be summarized as "Internet and more specifically digital platforms, have become a new public sphere (albeit private) to be monitored and punished with the indispensable complicity of police, armies and large corporations."

Although discussions focused on the digital world, issues such as web infrastructure, proprietary architecture of platforms, algorithmic logics that nurtured filter bubbles, viralization of certain content, dark ads, and microtargeting were left out of the debates. In other words, it was assumed that there was a big problem of disinformation, but it was treated following the logics that would have been applied to the journalistic sector of the nineteenth century without understanding that the logics, technologies, and mechanisms of digital capitalism were different. It was the equivalent of talking about the 2008 financial crisis without talking about financial deregulation, banking, or neoliberalism, with the added problem that, in this case, we were still counting on the large digital corporations to solve the problems that they, by design and by interest, had contributed to creating.

My participation focused on trying to show how digital capitalism had broken the rules of the game by explaining how the whole of our communications, our access to information, to indispensable services, passed through a handful of companies mostly protected and shielded in US jurisdictions. These were companies whose architecture was specially designed to viralize the contents that so concerned the parliamentarians—whose business model, based on the monopolistic control of data expropriated and exploited from users, clashed with individual and collective rights. As I said, little could be done against disinformation and fake news without entering fully into the regulatory debate on the digital architecture of large platforms and proposing public and citizen alternatives for communication and digital entertainment outside of a business model based on intensive surveillance, accumulation, and trade of data. How does one explain this to a group of parliamentarians generally disinclined—whether for ideological reasons or pragmatic criteria—to take measures against the irresponsible corporate power of the big technology companies? How does one warn them that paper laws are not enough to regulate the digital code? How can one make them understand that this is a problem that transcends ideologies—that this affects them as representatives of a country, but also as fathers, mothers, friends of vulnerable people? How does one convey to them that we could not expect solutions from notorious and recidivist corporate criminals? The very indifference of big tech to the common good made the task easier.

Around the same time, (former Facebook-Meta employee) Frances Haugen's first revelations had already begun to emerge in what would later become one of the biggest Meta-Facebook scandals. The *Wall Street Journal* revealed in October 2021 that the company had been conducting studies for three years to investigate the mental health effects of one of its most popular subsidiaries: Instagram.[52] The studies probed how a significant segment of its users, mostly female adolescents and tweens, exhibited symptoms of depression and anxiety, as well as suicidal and self-harming tendencies. Facebook's own research suggested that Instagram was significantly influencing the self-perception of women and girls. As the *Wall Street Journal* reported, "Two percent of teenage girls said that when they felt bad about their bodies, Instagram made them feel worse."[53] The same article also states that 40 percent of British teenage female Instagram users considered the app to be the source of their negative self-perception. This is consistent with recent academic contributions that point to the tremendous influence that social platforms have on the production and shaping of the self, and identity construction, in younger age groups. Despite knowing the pernicious effects of its application, Facebook executives decided to continue with business as usual. After all, the motive that induced millions of teenagers to enter a vicious cycle of social comparisons with harmful effects was the same one that made Instagram successful. Facebook was clearly putting profits before health, before safety, before the common good. Despite the evidence provided by Haugen, which revealed how the company's reckless behavior was causing enormous social suffering, no criminal proceedings were brought against the Silicon Valley company at the time. This is an example of the *crimes of the powerful*.

This concept, first coined by the Marxist criminologist Frank Pearce, can be defined as those behaviors that, although considered socially harmful, tend to fall outside the sphere of punitive power or, when they are recognised, are either rarely prosecuted or are punished in a lenient manner.[54] Is it out of malice, idleness, ineptitude, impossibility? The reasons are complex, but as Pearce and the long trail of critical studies that follow him demonstrate, the crimes of the powerful are not examples of *deviant behavior* but of functional acts, a necessary part of the capitalist social formation that requires the expropriation of surplus value, the destruction of nature, or systemic repression to guarantee its survival.

These harmful and damaging practices committed by "well-established private and/or public organizations in violation of the rights of workers, women, children, taxpayers, consumers" are not considered as deviant gestures, as outside the norm, but are perceived as a natural part of the system.[55] They are hidden by a dense ideological veil that justifies, legitimizes, and explains as normal what in other instances would be considered theft, fraud, kidnapping, or murder. In some cases, these practices are crude and obvious. For example, the so-called Start-up Nation, Israel,[56] boasts of its "smart solutions for riot control" such as the Cyclone drone or the "combat proven" Hermes UAVs that have ravaged Gaza.[57] Cyber(warfare) corporations such as Elbit Systems and Ispra profit from the export of military technologies that they have used as a particular laboratory for the indiscriminate repression of the Palestinian people.[58]

The crimes of the powerful commonly fall outside the reach of the state or are enforced only through civil and administrative sanctions, regardless of the magnitude of their socially harmful effects. Google, for example, has violated competition and antitrust laws in multiple jurisdictions with effective impunity. As the EU stated in 2019, Google's conduct "harmed competition and consumers, and stifled innovation."[59] Despite the proven offenses and the great social harm caused, no Google executive was charged, prosecuted, or investigated. However, it is fair to acknowledge that these incidents culminated in substantial fines being imposed on the corporation, amounting to over €8.8 billion. Although of course, these penalties only represent a fraction of their profits—accrued by the company through these same transgressions! *Misbehaving* is not only inexpensive for corporations, it is often deeply profitable.

Perhaps one of the main explanations for corporate impunity has to do with the way in which the field of criminal law itself has been designed and constructed. This is not the place to trace a genealogy of punitive law, punishment, and imprisonment or how these phenomena are directly linked to an ideology (liberalism), to a hegemony of class, gender, and race (white male owners of the means of production), and to an economic system (colonial capitalism). Others have brilliantly explained how law has been, and continues to be, a fundamental tool for reinforcing a brutal system of inequality that generally protects the interests of the powerful while punishing the poor.[60] As Soviet scholar Evgeny Pashukanis once put it, "Criminal justice in the bourgeois state is organized class terror, differing only in degree from the so-called emergency measures taken in civil war."[61] Criminal law, like any other expression of juridical form, is a social construction. In other words, it is the result of complex social, historical, economic, and cultural processes that change and shape the law in the image of society. Criminal law reflects hegemonic belief systems to the extent that it mimics and amplifies their hierarchical structures and the system of inequalities in which it is embedded. Crime has no real ontological basis; its perception and construction respond to philosophical contingencies and political agendas.[62] Although there is a multiplicity of contradictory perspectives, approaches, and opinions, the common agreement in critical criminology (inherited from an unorthodox view of Marxist legal analysis) is that criminal law favors those who are higher up in the scales of power and, therefore, evolves according to their interests. This is due to unbalanced and asymmetrical power relations. After all, companies, states, and individuals occupy very different positions when it comes to legislating, judging, and being judged. In other words, socio-legal structures are arenas of economic and political power with extraordinarily disproportionate subjective differences. A corporation with an army of well-paid lawyers and a direct connection to political power is not in the same position as an environmental organization. Paradoxically, the former can devastate a forest, pollute a river, or permanently destroy the ecosystem of a prairie and be considered not criminals but rather agents of progress; at the same time, the latter can be qualified as ecoterrorists just for protesting against this destruction.[63] Despite liberal dogma, we have never been equal before the law.

Critical criminology has attempted to explain why the criminal law is ill-equipped to deal with structural crime and social harms caused by the powerful, both private and public. Often the crimes of the powerful are not considered as serious as other types of crimes; they are justified or are treated leniently.[64] In other situations the serious social harm caused is not even recognized as a crime (it is illegal to steal an apple, but not to pillage an entire nation).[65] At other times, the crimes of the powerful may be subject to punishment, but the procedural barriers to achieving justice may be excessively costly, intricately entangled in legal processes, plagued by difficulties in obtaining critical information, or even dangerous for marginalized individuals or communities.[66] Consider for example the widespread use of automated and semiautomated risk assessment and crime prediction systems by law enforcement agencies in matters of national security such as terrorism or everyday criminality such as drug trafficking. As activist voices in places as different as New York, Barcelona, and Melbourne have pointed out, repressive institutions (public and private) are reluctant to explain and justify how human and technological elements capture, store, and process the data that feed the automated systems with which the police designate neighborhood shops in racialized territories as hot spots or signal the dangerousness of certain subjects.[67] Finally, as Edwin Sutherland argues, the powerful share a set of cultural, social, and political values.[68] A common framework of understanding unites white-collar criminals with those who legislate, judge, and police, making this group an indistinguishable gang. For Frank Pearce (and many others after him), this is nothing more than the expression of capitalist class hegemony, of ruling class solidarity. Some of the most palpable examples of pervasive social harm, often ending with state and corporate impunity, involve racism and patriarchy.

Safiya Umoja Noble, Ruha Benjamin, and Meredith Broussard, among other voices, have denounced how big tech algorithms reinforce patriarchal and racist social structures.[69] A palpable example of this phenomenon is found in automated resume processing, which has repeatedly prioritized traditional subjects of privilege while disproportionately affecting women, people from lower-income neighborhoods, and racialized communities. An example of how the classist, sexist, and racist ideology of corporations is automated. Similarly, several studies have shown that facial recognition technologies, widely used in the United States and other countries, tend to have higher error rates when identifying darker-skinned people (who are more frequently targeted by police interventions). This disparity in facial recognition accuracy has led to discriminatory consequences, such as misidentifications and false accusations that particularly affect racialized communities.[70] This is clearly not a technological problem of how well or how poorly a device can identify a face. The underlying issues are the structures of surveillance, oppression, and exploitation to which poor and racialized populations have been subjected and which the current algorithmic matrix extend. The data sets used to feed automated selection or surveillance processes are nothing more than a corporate and state glorification of racist, classist, and patriarchal arguments.

The harmful behavior of states and corporations is not limited to the passive design of racist technologies; it includes their active cooperation in organized terror campaigns against vulnerable populations. This is what critical criminology calls state-corporate crime. One of the earliest examples of state-corporate crime involving major technology companies took place at the height of fascism in Europe in the twentieth century. The respected computer giant IBM collaborated actively with the Nazi regime in the 1930s, building for it a tabulation and punch card system that would allow the classification and racial profiling of millions of subjects. According to the lawyer for Jewish survivors who sued IBM, "Hitler could not have so quickly and efficiently identified and rounded up Jews and other minorities, used them as slave laborers and ultimately exterminated them, without IBM's assistance."[71]

This was also one of the first examples of racist treatment of computerized data, something that has so far had no consequences for the company.[72] As I will show, large tech companies have collaborated with immigration agencies around the world to design, manage, and implement technologies that have resulted in serious human rights violations, such as deportations and family separation. Palantir, for example, has worked closely with US law enforcement agencies, providing them with a wide range of predictive policing tools. These technologies, as has been denounced in a multitude of places, are often biased against racialized communities by associating them with gangs, mafias, and drug trafficking—a technological reaffirmation of the discourses of the extreme right.[73]

Another example of the crimes of digital capitalism that often elude the punitive sphere of the state has to do with environmental destruction. Big tech has often been labeled as a clean, green, and socially conscious industry and as the only alternative to the climate emergency. However, behind the veil of creativity and abstract cyberexistence, the digital economy depends on a material infrastructure that is no different in its impact from other industries with less ecological pedigree. Corporations such as Google or Apple depend on a long and varied supply chain that extends from mining, fundamental to the material existence of wiring and the material production of goods, to water, wind, and energy sources for the operation of their data centers and fossil fuels for the global distribution of consumer goods.[74] Some estimates indicate that the demand for the so-called energy transition minerals will increase exponentially in the coming years (about forty times in the case of lithium). As with other resources, most of these minerals are located in Indigenous territories. According to a recent estimate, 52 percent of all mineral reserves needed for the green and digital transition are located in Indigenous territories, a figure that rises to more than 70 percent for "critical" materials such as lithium.[75] Despite the fact that different international and national legal instruments recognize Indigenous sovereignty over their territories and the right of these peoples to be consulted and give their consent for the intervention and exploitation of their lands, the fact is that it is the capitalist logic that prevails. Countries such as Argentina, Chile, Canada, Brazil, Australia, or Peru qualify new mining, energy, or infrastructure projects as critical, fundamental for national security, reaffirming in the same movement both the colonial state regime over Indigenous peoples and extractivist capitalism.[76] Indigenous peoples and peasants, historically dispossessed and impoverished, are offered few options: accept the supposed (and short-term) economic advantages offered by mining industries, crumbs of the gigantic profits obtained by BHL, Glencore, or Rio Tinto, or be considered as ecoterrorists for defending their ancestral territories.[77]

The environmental impacts associated with digital capitalism also extend to logistics operations related to service provision. The "disaster" of recent one-day shipping practices such as those offered by Amazon Prime amplify the already exaggerated environmental impact of compulsive consumption.[78] An additional example of ecological crime is the growing accumulation of highly polluting e-waste. As noted by Interpol in 2019, the e-waste industry stood out as one of the largest and fastest growing illegal industries globally. It is estimated that around fifty million tons of e-waste is generated annually, with a significant portion of this being processed and transported outside official legal channels to particularly marginalized regions of the Global South such as in Agbogbloshie (Ghana) and Nairobi (Kenya), exposing affected local communities to environmental degradation and hazardous carcinogenic substances.[79] Much of this e-waste is the result of deliberate corporate strategies. For example, large technology companies have pushed against the right to repair, imposing a high financial burden on those who are willing to repair rather than buy new devices. In the same vein, big tech has resisted timid regulatory moves against planned obsolescence that artificially shortens the useful life of the goods they produce.[80]

But perhaps one of the most blatant examples of the routinization and legalization of social harm lies precisely in the very existence of corporations. After all, these artificial entities are at the epicenter of many of the current crimes of the powerful. This legal fiction took shape during the dawn of bourgeois hegemony, around the sixteenth century, in what is now Western Europe. It was a period characterized by colonial expansion, the brutal dispossession of entire territories, human trafficking, genocide, epistemicide, slavery, and institutionalized rape culture. As profitable as it was, the colonial enterprise was extraordinarily expensive, requiring arms, munitions, ships, payments, and so on. The colonial powers lacked the funds to satisfy their imperial ambitions, and the early capitalists were unwilling to risk their wealth in uncertain ventures. The solution was found in the joint-stock company, a form of public-private partnership that would provide colonial enterprises with the necessary financing while guaranteeing the bourgeoisie economic and legal protection. The powerful felt secure. The corporation, while carrying rights similar to those of natural persons, was not criminally liable.[81] In other words, the ruling class deformed legality to create a shield of irresponsibility around their crimes. Grietje Baars has brilliantly described corporations as "capital personified," a "structure of irresponsibility" created to ensure "corporate impunity."[82] Thus leviathans such as the East India Company departed from the major colonial ports of Europe with the dual aim of asserting both imperial and capitalist hegemony. This was the dawn of the global state-corporate terrorist campaign against the peoples of the world. The same structure of impunity and corporate irresponsibility that has enabled massive devastation in every corner of the Earth continues to operate today. Corporations with power and influence greater than that of most states have manipulated, violated, or pressured public institutions to shape national and international laws to their whim, allowing them to exploit the working class under new-old formulas of precarization, extract data massively and indiscriminately, and profit from the privatization of public goods and services.

The geology of the social damage of digital capitalism is complex. It ranges from modern slavery in the context of mineral extraction to Taylorist exploitation in Chinese hardware production industries. It shifts to machinic dispossession in automated warehouses, algorithmic discrimination in welfare systems, and the deployment of policing technologies by private and public actors. Although we all suffer the consequences of a digitized global structure of inequality, the resulting social damage is not equally distributed. The crimes of digital capitalism are, by nature, an inherent and organizational part of an imperialist phenomenon that Couldry and Mejias define as "data colonialism."[83] Hegemonic digital corporations are often headquartered in countries in the Global North, predominantly the United States, while their victims are disproportionately located in the Global South. Moreover, this structure of inequality is intertwined with prior forms of racial, class, and gender oppression, disproportionately affecting the most vulnerable in the Global North. Confronting the crimes of digital capitalism, which affect us both individually and collectively, requires going beyond failed tough-on-crime strategies, regardless of whether these might be directed against the powerful. We cannot fall into the temptation of a populist punitivism that would be nothing more than the reaffirmation of the deep logic of prison ideologies. We need new approaches, methods, and institutions of the commons capable of enforcing concepts of justice that lead to a real and profound transformation that shakes the causes of social damage. Thinking about the crimes of digital capitalism is an exercise aimed not at reaffirming mechanisms for punishment but at producing the necessary tools to abolish the conditions that make them possible.

Approach and Structure of the Book

There is no doubt that the dire and far-reaching impacts of digital capitalism have captured the attention of activist and academic researchers. In recent years, a number of excellent works have emerged exploring from multiple angles, disciplines, and approaches the political economy, structures, technologies, mechanisms, policies, and, of course, negative consequences linked to the digital age. For example, the works of Christian Fuchs, Nick Srnicek, and Jathan Sadowski have helped us understand the political economy of socio-technical phenomena such as smart cities or large digital platforms. Authors such as Shoshana Zuboff and Mark Andrejevic have explained the logics of surveillance on which digital capitalism is built.[84] Researchers and programmers such as Safiya Umoja Noble and Timnit Gebru have exposed algorithmic racism and its corporate concealment. Authors such as Rashida Richardson, Sarah Brayne, Ana Muñiz, and Petra Molnar have highlighted the radical change in punitive institutions, such as the police, and the control of frontiers in their encounter with new technologies of biometric control, massive and indiscriminate surveillance, and automated systems.[85] Voices such as Nick Couldry, Ulises Mejias, and Tahu Kukutai have brought us closer to aspects such as data colonialism and Indigenous digital sovereignties.[86] Avant-garde authors such as Meredith Whittaker, Jathan Sadowski, Salomé Viljoen, and Ekaitz Cancela invite us to think about alternative futures and digital utopias beyond existing schemes.[87] These are works that, despite their diversity of approaches, methods, and theories, have an important trait in common: identifying how power—whether public or private—uses digital technologies to entrench a social structure of domination that benefits the few while harming the many. Despite the wealth of critical contributions, there is no work that addresses the criminal structure of impunity fostered by digital capitalism along the supply chain, nor work that systematically analyzes its causes and mechanisms and the actors responsible for it. And it is precisely to fill this gap that I have set out to address in this book.

To carry it out, I have immersed myself in an exhaustive review of academic works, institutional reports, journalistic articles, and regulations. I have created work of review and analysis that navigates disciplines such as criminology, sociology, political economy, and technology studies, while maintaining the critical prism offered by decolonial and Marxist theories. This theoretical review is accompanied by numerous case studies and field research, which exemplify how the rise of algorithmic racism (as in the Netherlands), green and digital capitalism (in Chile or Extremadura), and the platforming of education (in Australia) are palpable realities affecting millions of people around the world. The book you hold in your hands is therefore a work situated in the literature of the crimes of the powerful. In it, I take an approach that over eight chapters has allowed me to explore the massive social damage caused by large technology companies and states, considering it not as deviance but as criminal strategies necessary for the existence of digital capitalism.

In this introductory chapter I have provided an exploration of the theoretical foundations from which I will analyze the crimes of digital capitalism. It is an overview, a window into the content to come. In the first chapter I analyze how the digitization of the neoliberal state project is giving rise to socially harmful forms of algorithmic governmentality, which are contributing to shoring up old structures of inequality, racism, and classism while fueling new forms of violence and discrimination. To this end, I unravel two interconnected phenomena of what I term *digitized racial neoliberalism*. First, I explore the ideological structure of racial neoliberalism and its articulation with the rise of far-right discourses and narratives at the center of state policies in much of the world. Through the study of a well-known case of technologized structural racism that took place in the Netherlands, I explain how these ideological foundations are used to build powerful information gathering and processing apparatuses, which, operationalized through automated decision-making systems, seek to punish, monitor, and dispossess populations classified as dangerous, lazy, or undeserving of aid.

In the second chapter I analyze the digital colonization of the educational sector, paying attention to two fundamental dimensions: the political economy of the educational technology (EdTech) sector and the consequences it is having on both students and educational professionals. Far from posing it as a disruptive phenomenon, I expose the digital corporate assault of face-to-face and virtual educational environments as the acceleration of the neoliberal privatization agenda of the 1980s. Under this situated framework, I scrutinize how digital giants such as Google are taking control of the new educational infrastructures and redefining pedagogies and the role of educators according to the values of datafication, quantification, surveillance, and digital capitalism.

In the third chapter I present a radical definition of cyberwar, highlighting its historical intertwining with processes of dispossession. In the first part of the chapter, I discuss the origins and limitations of the concept from a critical and Marxist perspective, proposing a new heterodox definition, which, fleeing from state-centric parameters, places the concept of cyberwar in a global context of dispossession, colonialism, and class struggle. Subsequently, and relying on a new wave of critical authors, I analyze different aspects of the progressive inclusion of military technologies in the civilian sphere, such as border surveillance, security, and control of political dissidence. In doing so, I intend to draw attention to the accelerated growth of the industrial-securitarian-military-academic complex and the dire consequences this is having for the population, especially the poorest and most racialized groups, who are increasingly seen by states and corporations as enemies to be defeated.

In the fourth chapter, I land on one of the most perceptible and perhaps least well-known territories of digital capitalism: the algorithmic exploitation of workers. As a starting point, I dive into the thorny debates that for much of the twenty-first century have sought to explain the fluctuating relationship between capital, labor, and the means of production: are Deliveroo, Amazon, and Uber local software companies, or giant global service employers? To answer this question, I pay attention not only to the evident technological change and the emergence of new models of extraction, exploitation, and surveillance but also to the increasing racialization of broad sectors. This is a theoretical debate neccesary to understand the nature of the conflict between racial capitalism and labor in a world where code is at once a means of production, manager, and labor law for millions of workers.

In the fifth and sixth chapters, I dive into one of the most worrying, thrilling, and relevant debates for the future of the planet: the socio-environmental impact of digital capitalism. Given the magnitude and importance of these discussions, I have divided the analysis into two chapters, which, although closely connected, address distinct issues. In the fifth chapter I dissect the hegemonic narrative that describes the so-called green and digital transition as necessary, environmentally friendly, and without tangible consequences. To this end, and drawing on anti-imperialist and decolonial reflections, I examine the materiality of the digital economy, exposing both the capitalist dogmas and interests that justify it and its insatiable need for natural resources. I then offer a look at an example of extractivist legal infrastructure designed to regulate, normalize, and naturalize a new wave of environmental and social devastation. In chapter six I look at two deeply interdependent questions: What kind of ecosocial impacts are being caused by the routine extractive operations imposed by digital capitalism? What kind of discursive, legal, and political mechanisms are available to address these impacts? In order to try to answer these questions, I dive into the recent debates on the criminalization of ecocide crimes from both theoretical and practical perspectives. Avoiding the usual channels, I approach this reflection from a critical, decolonial, and abolitionist perspective, an approach that may allow us to understand and think of collective response mechanisms against the socio-environmental destruction of ancestral territories that, as in the case of El Salar de Atacama, are being devastated under the protection of the green and digital transition.

With the conclustion I pursue a twofold objective. First, I seek to offer succinctly the fundamental concepts and keys to what I have called the crimes of digital capitalism. Second, I intend to confront what at first sight might appear to be one of the most glaring contradictions and limitations of my own arguments: to what extent can an approach grounded in criminology, and thus based on state structures rooted in capitalist and colonial oppression, effectively address the structural harm perpetrated by states and corporations? Nevertheless, there is no reason to worry. I do not intend to conclude the book with a mere summary or an idealized and abstract deployment of critical theory. Instead, I have reserved for the end the analysis of one of the most socially damaging digital corporations: Meta-Facebook. With the analysis of this case, I intend to go beyond critique and delve into potential policy proposals. That is, imaginative and collaborative ways to build alternatives to (digital) capitalism. In other words, with the conclusion, I do not propose an end or a reflection on what is raised in the book, but an opening, a door to the exploration of solutions, of alternatives, perhaps of renunciation and escapes, perhaps of transformations, and, in any case, of paths toward a world that is more necessary than impossible.

1

The Digitization of State Racism

On January 15, 2021, major European newspapers reported a surprising news story. The Dutch government resigned en bloc as a result of a social benefits scandal. Around twenty thousand families were falsely accused by the Dutch tax authorities of fraudulently claiming benefits. Just over ten thousand of these families were forced to repay the benefits (in amounts totaling more than €30,000) while their access to other benefits was canceled. Thousands of people were doomed to ruin, deprivation, and eviction. Families were singled out and pushed to stress and social stigma, which ended, in some cases, in divorce and suicide.[1]

"These are tens of thousands of families crushed under the wheels of the state," Dutch neoconservative prime minister Mark Rutte told the press.[2] What the Dutch politician forgot to mention is that the families were mostly living in a handful of disadvantaged neighborhoods, disproportionately from non-European immigrant backgrounds. To some Dutch institutions, it was a technological error, a wrong algorithm. To others, it was human error, failures in the supervision of the institutional machinery. According to Rutter's own cabinet, it was a violation of the rule of law. What was never made clear was that this gross and massive disenfranchisement of racialized families was not a mistake but the predictable result of the racial neoliberalism advocated by conservatives reluctant to pursue the rich with the same zeal.[3] And, of course, it was the result of the bundle of new algorithmic technologies with which they were putting it into practice.[4]

Since 2014, the Dutch Tax Agency has been using an algorithmic tool called Systeem Risico Indicatie (SyRI or, in English, System Risk Indication). SyRI is not a magical artificial intelligence capable of predicting the future. Nor does it identify people who have violated laws. Like other risk assessment technologies, it is a system that, based on human-coded statistical criteria, determines the probability of a subject doing (or failing to do) something, for example, committing tax fraud. SyRI examined and compared economic data from various government databases—from fines, taxes, debts, penalties, property, benefits, and subsidies—trying to establish correlations that would indicate fraudulent use of subsidies.[5] To this supposedly neutral and objective information, in order to delimit the probability of committing fraud, other variables were added that had little or no relevance in determining the "risk" of committing infractions. Not being a Dutch citizen or having dual citizenship became in the eyes of the Dutch algorithmic administration a symptom of dangerousness, an indicator of risk.[6] This is a clear and simple example of how racist prejudices are installed in technological logics. For some, the solution to this type of algorithmic bias inscribed in the architecture of the systems is to include the human in the technological loop, for example, by establishing supervision mechanisms. This point of view is based on a premise that is as simple as it is wrong. What is known in the industry as discrimination or algorithmic bias is a glitch, a system failure, a technical error that can be corrected with adequate personnel and resources, a clear regulatory framework, and a good professional education in values.[7]

However, the Dutch system was not fully automated. According to national and EU laws any automated decision with potential impact must include human supervision. In other words, Dutch officials had to personally review high-risk applications flagged by the system. The Netherlands is a leading country in indicators of development, wealth, and tolerance. What could have gone wrong, then, when the human mechanisms that should have prevented the damage caused by a biased automated decision-making system were in place?

As recent contributions show, human supervision of automated decisions often becomes a symbolic gesture without content in which people mechanically confirm and validate the algorithmic decision.[8] This is not incompetence or sloppiness; it is a deliberate process, the result of a three-pronged strategy based on control, deprofessionalization of the workforce, and outsourcing of public services. Among others, Virginia Eubanks has shown how under the discourse of algorithmic efficiency, neoliberal administrations seek to cut social spending. This takes place both in the labor dimension—automating part of the work—and in the availabilty of aid, imposing barriers and forms of technological punishment to those who supposedly take advantage of social resources.[9] Public and social service workers often find themselves having to respond to an increasing number of cases with fewer resources and personnel. Systems such as SyRI are offered to the public and to the workers themselves as the techno-solutionism to all problems, as a seemingly neutral and infallible aid, capable of assessing cases in seconds, to which overburdened civil servants flock and to which they rarely object, thereby legitimizing both their own deprofessionalization and the potential negative consequences of the system. But, in addition to this neoliberal privatizing dynamic common to a large part of the countries of the global North, there is another dimension, that of structural racism, a phenomenon that is as invisible and hidden as it is widespread.

SyRI singled out certain subjects for their "dangerousness" but did not report the reasons that had led it to classify one or another application as "risky." The workers were presented with a document that they had to review. They could doubt the machine and work hard to grant aid or they could doubt the person and look for signs of fraud. They opted for the latter. A mixture of blind faith in technology and structural racism set the stage for disaster. Throughout the years in which the SyRI system was in operation, public officials justified technological racism with bureaucratic racism by searching thoroughly for human errors in the applications flagged by the system, such as by marking as gross negligence applications with errors and minor faults such as delays and omissions in signatures.[10] In other words, the digitization of state racism in the Netherlands was not the consequence of a hallucinatory, cold, or unethical technology. It was carried out under human supervision.

Bureaucratic racism was not limited to this process but extended throughout all steps of the procedure. As the Dutch civil rights platform Bescherming Burgerrechten denounced in 2018, public institutions were under no obligation to inform affected families.[11] When they were informed, in many cases, it was done by indirect means such as bulletins and in vague terms that did not describe what kind of information had been collected nor how it had been used to support algorithmic decisions. Added to the algorithmic opacity was the usual bureaucratic black box creating a perverse and damaging synergy. Families found themselves in a position of helplessness, facing misinformation and the impossibility of accessing proper legal procedures. All this while having to navigate a system that was supposed to safeguard their interests but instead had declared them enemies.[12]

Finally, after years of organizing and mobilizing, in February 2020, a court in The Hague declared the SyRI system illegal. To the surprise of many, this ruling did not recognize the widespread violation of civil, economic, and political rights. It was justified on the basis of the violation of Article 8 of the European Convention on Human Rights (ECHR), which safeguards the right to privacy and family life, as well as the inviolability of the home and correspondence. As the judgment recognized, the Dutch Tax Administration misused the various records at its disposal. Specifically, it used the country of origin of benefit claimants as an indicator of their risk profile.[13] The court found that SyRI's implementation failed to reconcile the interest in fraud detection and the human right to privacy and declared the system unlawful. At first glance, this might appear to be a positive outcome—after all, the courts were reacting by condemning the administration. But a closer analysis reveals that, in reality, this was nothing more than a false move. It was a gesture by which one of the biggest scandals of structural and institutional racism in recent Holland was reduced to the category of a violation of privacy. It was only thanks to pressure from families and activists that the Dutch Parliament appointed a commission of inquiry in July 2020. This was followed by parliamentary hearings with a strong media impact which resulted in a political shake-up.[14] What started as a local episode of data mismanagement became the first case in history in which a government fell as a result of an automated decision system.

The scandal revealed a brutal reality in which the most vulnerable and fragile sectors of a population are monitored, singled out, criminalized, and illegally dispossessed of their rights by the institutional apparatus of a wealthy State using the latest technological advances. But this is not an isolated episode. Governments around the world are turning their eyes to the promises of the big tech companies to provide technical solutions to complex and structural problems. A good example of this is India's creation of the Aadhaar system, a vast framework or infrastructure comprising state surveillance technologies designed to collect biometric data on the entire population. It is intended to serve as the main access point for a wide range of public and private services, such as identification, payments, commercial transactions, and employment services.[15] Barcelona City Council is also developing an automated system to determine eligibility for specific services to its welfare "clients." New Zealand, a pioneer in the digitization of its welfare structure, uses automated systems to calculate the amount of welfare benefits to which each individual may be entitled.[16]

Australia has infamously used algorithmic technologies to prevent fraud, resulting in massive welfare surveillance scandals.[17] The United States, meanwhile, has spearheaded a digital welfare dystopia, where large corporations are codifying conditions of oppression and algorithmic surveillance of the most vulnerable.[18] Whether they are developing their own internal automated welfare decision systems or entrusting it to private corporations, governments around the world and across the political spectrum are veering toward the solutionist mantra: Tech will save us. The technocratic, solutionist, and centralist political ideology championed by Silicon Valley has found its way into the various iterations of the neoliberal state project, giving rise to what academics and activists have already dubbed the "digital welfare state." In the words of Philip Alston, "Our systems of social protection and assistance are increasingly driven by digital data and technologies that are used to automate, predict, identify, surveil, detect, target and punish."[19]

In this chapter, I will explain how the ideology and political economy of racial neoliberalism coupled with the socio-technical advances of the digital age are leading to a socially harmful form of algorithmic governance and a technocratic and solutionist project that reinforces inequality and xenophobia while threatening fundamental freedoms. To explain this process that imbricates state racism, neoliberalism, and algorithmic oppression, I will focus my attention on the SyRI system variable (whether the user had Dutch citizenship) that ended up leading the Dutch government to resign in 2021. As I will demonstrate below, the apparent simplicity of the variable hides behind it a punitive and criminalizing political agenda guided by a dangerously rising model of digitized racial neoliberalism.

Political Tools of Algorithmic Oppression

Automated decision-making systems, such as SyRI used by the Dutch authorities, are not just pieces of software but a set of social, political, and digital technologies. They encompass the people who design and operate the system, the users, the data sets, the computational networks that connect the elements, the institutional architecture in which they are embedded, and the broader sociopolitical structure of the place where they are built.[20] Scholar and activist Rashida Richardson offers a rich and extensive definition of automated decision-making systems as "any tool, software, system, process, function, program, method, model, and/or formula designed with or using computation to automate, analyze, aid, augment, and/or replace government decisions, judgments, and/or policy implementation. Automated decision systems impact opportunities, access, liberties, safety, rights, needs, behavior, residence, and/or status by prediction, scoring, analyzing, classifying, demarcating, recommending, allocating, listing, ranking, tracking, mapping, optimizing, imputing, inferring, labeling, identifying, clustering, excluding, simulating, modeling, assessing, merging, processing, aggregating, and/or calculating."[21]

Renowned experts in algorithmic law and regulation Mireille Hildebrandt and Karen Yeung have separated automated decision systems into two main categories: reactive/code-driven systems and preventive/data-driven systems.[22] Reactive/code-driven systems respond to simple control structures—give access, open passage—such as those you typically find in passwords, locks, or traffic lights. They are based on predefined standards that can potentially be extended to form complex decision trees. Despite their size, they operate in a deterministic and predictable manner, following established rules, regardless of their complexity or the nature of their technological components. In contrast, preventive/data-driven systems are designed to anticipate risks. To this end, they integrate information extraction strategies, statistical tools, and (neo) actuarial methods to develop (supposedly) predictive technologies.[23]

The progression of these preventive and predictive capabilities gives public and private owners of these systems the ability to collect, process, and share vast amounts of data from numerous sensors and sources (or to acquire it through data brokers). But it's not just about information management, it's about executive power. These systems can be configured in such a way that, when faced with specific circumstances, they can provide specific responses. For example, when income received exceeds certain undeclared thresholds, the system could issue a warning letter or apply a sanction. These are not hypothetical models; these technologies are widely applied both in the private sphere (insurance, mortgages, contracting) and in the public sphere, especially in the welfare and criminal justice systems. This is already causing a profound social impact as illustrated by the SyRI case in the Netherlands and, as another example, by the so-called Robodebt scandal in Australia.

In 2016, it was revealed that the Australian government, through its digital platform Centrelink, implemented an automated debt recovery program. By cross-referencing data from tax and welfare services, the system generated debt letters to hundreds of thousands of people for an estimated total amount of about A$1.2 billion.[24] Subsequently, a court ruling found that the neoliberal algorithm generated at least 470,000 *incorrect* debts.[25] Despite the gigantic dimensions of the case and its enormous social repercussions, it was not until the development of a Royal Commission of Inquiry (the final results of which were published in 2023) that some politicians began to reluctantly assume responsibility for what Prime Minister Anthony Albanese called a "gross betrayal and a human tragedy."[26] This chain of more or less forced resignations reveals the real political extension of systems that are, in principle, considered to be technical.

Kathryn Campbell, secretary of the Department of Human Services during the implementation of the scheme, was suspended from her position in the defense department in 2023. Alan Tudge, minister of Human Services between 2016 and 2017, initially rejected allegations of abuse of power in his media handling of the scheme eventually resigned from his positions. Christian Porter, minister of Social Services between 2015 and 2017, and Stuart Robert, head of Robodebt in 2019, also accepted their complicity in this massive crime against Australians. However, the one who was the political architect of the case, the minister of Social Services between December 2014 and September 2015, and later prime minister from 2018 to 2022, Scott Morrison, denied his responsibility while attacking the Commission of Inquiry which he accused of political persecution.[27] For the neoliberal politician, the Robodebt case was the result of a chain of errors, misunderstandings, and technical failures. Nevertheless, the commission's report affirmed the political character of Robodebt, asserting that the system "was precisely responsive to the policy agenda that had been communicated to the Social Security portfolio departments, both in private meetings and in the public sphere."[28]

Can the Digital Subaltern Speak?

There is an extensive literature, in what could almost be considered a scientific field in its own right, analyzing the close relationship between statistics, the techniques of quantification of the self, and the rise of the modern disciplinary state—a genealogy that must be studied in detail to understand the hows and whys of contemporary algorithmic racism. Michel Foucault laid the foundations for understanding the dynamics between scientific knowledge and the power of management over one's own life. Throughout his work, we find evidence of how the sciences and their methods were used to classify, control, and subjugate human groups, thus revealing the intrinsic interconnection between statistics, calculation, and power structures. Following this line, Tukufu Zuberi demonstrated how socio-statistical tools are not mere neutral instruments of measurement but armaments used in the service of white racial supremacy, technologies of power historically used for the control of inferiorized populations, if not for their extermination.[29] Santiago Castro-Gómez has shifted the Foucauldian analytical framework to Latin America. In his indispensable work "La Hybris del Punto Cero," he examined how science played a fundamental role in the articulation of the binomial colonial power and racism, mobilizing Eurocentric epistemologies that marginalizes Indigenous and Black forms of knowledge and strengthened the imperial regime.[30] In this line, Umamaheswaran Kalpagam revealed how statistical and calculating tools became instruments of British colonial rule in India, contributing to the creation and perpetuation of imperial power structures.[31] For her part, Caitlin Rosenthal has investigated the role of calculus and accounting on US slave plantations, showing how accounting techniques were essential to the perpetuation of this regime of exploitation and control which it reinforced.[32] More recently, Wendy Chun has explored how current practices of data collection, correlation, and analysis are used to amplify prejudice and reinforce the regime of racial capitalist inequality.[33]

But the datafication of subjects is not just a way of managing the distribution of what exists, it is a productive, generative technology. As Chun points out, correlation processes driven by machine learning not only classify people, but also generate them. These scientific mechanisms of racial production are rooted in centuries of statistical advances, probability models, and racist-capitalist (pseudo)sciences such as eugenics.[34] In his influential work "The Taming of Chance," Ian Hacking describes how the expansive Prussian state designed one of the most formidable statistical apparatuses of the second half of the nineteenth century.[35] One of its most disturbing results was the development of mathematical tools designed to analyze and frame the "Jewish question" as a social problem. In tune with other contemporary discourses pointing to the decline of the superior races, the German intellectual elite perceived the physical and moral decline of Germany and set out to investigate the causes of such a dismal problem. The scientific elite of the time believed they could find the answer in the significant influx of Jewish immigrants from the Austro-Hungarian Empire and Russia to the East Prussian regions. Under the guidance of the enlightened Prussian scientific leadership, the Jews in Germany were systematically enumerated, racialized, and differentiated from the "true" German population of the Empire. These immigrants were produced as a race apart, as a seemingly dangerous population to be controlled and governed.[36]

At first glance it might seem that the algorithmic technologies behind systems such as SyRI or Robodebt bear little resemblance to previous control techniques such as the imperial science used to produce population censuses or eugenic statistics. However, despite the level of development in data collecting, processing, and cross-referencing technology, fundamental political questions about the processes of subject datafication remain: Who dataficates whom? Who decides how and for what purpose data are processed? Who asks which questions about whom using which technologies? Marginalized and racialized populations had no voice within the regime of colonial census domination imposed on them, just as they do not have a voice today in the modern processes of algorithmic numerical governmentality with which they are managed. Researching, measuring data, and interpreting results has been and continues to be a way of doing politics that inevitably requires an understanding of justice.

As I will explain, the contemporary predictive tools used to create "virtual borders" and, in the fight against terrorism, the "weapons of mathematical destruction" that Cathy O'Neil talks about are no better than the proto-Nazi apparatus designed in the Prussian Empire.[37] In other words, they are glorified socio-technical devices used to validate and legitimize a particular political stance. As numerous journalists, sociologists, and philosophers of technology have demonstrated,[38] there is no such thing as neutral technology. The very code in which these socio-technical systems are written embodies the politics of their time, which today amounts to neoliberal racism. But what does this concept mean and in what ways can it be related to algorithmic systems such as SyRI or Robodebt? In the following sections, I will focus on these thorny questions that furrow the political matrix of contemporary algorithmic oppression.

Neoliberal Racism: Beyond Discrimination

The concept of neoliberalism is a complex and polymorphous label used to denote an ideology, an economic system, a legal doctrine, a set of political and business ideas, a sociological theory, and a system of knowledge production, among many other things.[39] French criminologist Loïc Wacquant challenged the elusive nature of the concept by outlining a sociological characterization of it. For Wacquant, neoliberalism would imply not only a set of market-oriented policies but the articulation of four interdependent logics: the positioning of the market as the "epistemological truth" necessary to coordinate the provision of goods; the replacement of social benefits by a quasi-contractual form of workfare; the emergence of an intrusive penal apparatus aimed at disciplining and governing "the precarious fraction of the industrial proletariat"; and finally the proclamation of state and corporate irresponsibility in tandem with the glorification of individual responsibility. Wacquant, drawing on Pierre Bourdieu, pointed to the continuity between the social and penal sides of the Leviathan through multiple strategies of surveillance, discipline, reward, and repression, executed by a network of public and private agents such as police, social workers, businesses, teachers, and politicians.[40] While sophisticated, Wacquant's analysis fails to capture in its fullness the density and importance of the categories of race and gender in the punitive power structures that enable the functioning of the neoliberal regime.

Drawing from Marxism and Black feminism, abolitionist activists and scholars Angela Davis and Ruth Wilson Gilmore explain how understanding the racist logic of contemporary capitalism requires a close analysis of the historical structures of state violence on which it depends. The thinking of both authors was marked by the shocking increase in the prison population in the United States between 1970 and 2000, a pivotal period in the development not only of policies of mass incarceration, but of the development and maturation of the most brutal and unbridled political neoliberalism. In this period, the number of people incarcerated in the United States rose from just over three hundred thousand to more than two million, a massive legalized kidnapping at the hands of the state that disproportionately impacted Black, Hispanic, and Indigenous communities. This punitive wave was articulated in a double movement: on the one hand, extensive cuts in social services, on the other, the multiplication of repressive and surveillance apparatuses.[41] For Davis, the prison industrial complex responsible for mass incarceration overflowed its repressive function, consolidating itself as a mechanism of governance and social control over racialized and subalternized populations and asserting itself as a regime of violence and premature death that sought to perpetuate the status quo of white supremacy over other communities. Following the path set by Davis, Gilmore delved into the socio-spatial distribution of violence and inequality in what she called "geographies of racialized power." Gilmore introduced the concept of "organized neglect," which explains the deliberate neglect of certain communities and areas by the state that are deemed undeserving of investment in education, industry, health, or social benefits. This stigmatizing neglect degraded populations and was followed by a regime of surveillance and incarceration. In fact, for Gilmore, punitive systems and networks of power transcend prison walls. Prisons are part of a system to which they contribute their regime of violence by maintaining historical regimes of dispossession and racialized accumulation. This phenomenon was not unique to the United States but, with different variations, was echoed in countries such as the United Kingdom, Australia, Spain, and New Zealand. Even countries with relatively low incarceration rates such as the Netherlands increased their prison populations while multiplying their punitive and criminalization strategies. It is not difficult to understand why both Davis and Gilmore focused their attention on prison institutions and the role they played as articulators of the political economy (and punishments) of their context.

As highlighted by Gilmore as well as a plethora of authors, including Arun Kundnani, neoliberalism was not built in a vacuum.[42] It rests on racist Eurocentric structures of domination, which have defined and shaped the pillars of the current social order. In this context, the works of Dylan Rodríguez, who traces the genealogy of current US racist multiculturalism to a permanent state of domestic and international regime-war,[43] as well as the contributions of Aileen Moreton-Robinson regarding the overlap between colonialism, Indigenous dispossession, and the imposition of a patriarchal regime of white oppression in Australia,[44] offer fundamental keys to understanding the web of power that underlies the aforementioned mass policing and incarceration. Both theorists illuminate the ways in which neoliberalism and coloniality converge, exacerbating the exploitation and oppression of racialized and Indigenous communities and contributing to the construction of a punitive system that reflects and perpetuates structural dynamics of dispossession and oppression. For their part, Latin American decolonial theorists such as Aníbal Quijano, Ramón Grosfoguel, and Silvia Rivera Cusicanqui have highlighted how the assemblage of colonial techniques of domination deployed by the European powers in the sixteenth and seventeenth centuries did not come to an end after the formal independences of the metropolises but remained active in the systems of power and thought ordering the political-racial reality of the new states.[45] Quijano has called this phenomenon the coloniality of power and knowledge, which would go on to configure the material and mental conditions of exploitation of what would later become global capitalism.[46] The global hierarchization of the labor market, the unequal distribution of wealth, the racialized distribution of datafication processes and the epistemic hegemony of the institutions of the Global North over their counterparts in the Global South are just some of the endless examples of the persistence of such structures.

But it was the Black Marxist school incarnated by authors such as W. E. B. Du Bois and Cedric Robinson that enhanced the understanding of this structure of oppression as racial capitalism. Robinson stressed that the racialized and repressive nature of American labor relations, defined by the color line, never ended with the abolition of slavery but persisted as a racial ordering mechanism within the white supremacist economic regime.[47] Historian Gerald Horne demonstrated how the epistemic and institutional roots of US architecture were designed to reinforce a system of domination based on the dispossession of Indigenous lands and the exploitation of subaltern labor.[48] The same can be said of the legal infrastructure. Other researchers have shown that analogous technologies have operated in both colonial states and metropolises.[49] For example,

Juan Tauri and Ngati Porou have detailed how the punitive welfare state of Aotearoa (Maori for New Zealand) is incarcerating the Maori population in similar proportions to the hyperracialized US Criminal Justice System.[50] Similarly, Amanda Porter has highlighted the recolonization strategies deployed by punitive welfare apparatuses on the Australian Aboriginal and Torres Strait population.[51] Robert Shilliam has demonstrated the way in which neoliberal racism intersects with postimperial melancholia in settings such as the United Kingdom, Spain, and France.[52] The legal regime of racial capitalism with its plethora of laws, codes, and ordinances, designed to sanction "racialized accumulation by dispossession,"[53] persists today in the digital sphere, through different strategies of algorithmic racism, something Ruha Benjamin has defined as the "New Jim Code."[54]

The exclusion of people from social benefits by the Dutch administration was not a mistake or a technical failure. It was a deliberately calculated strategy linked to a defining feature of neoliberal racism: the differentiation between those deemed to deserve the social protection of the state and those who should be subject to surveillance, punishment, control, and violence. The Dutch parliamentary inquiry cited above revealed that the choice of where and on whom to deploy the socio-technical systems used to detect fraud followed political criteria founded on racism and prejudice. These systems disproportionately targeted people from former colonies (mainly the Dutch Antilles) and immigrants from Muslim-majority countries. This discriminatory practice reflects a latent eugenic-racist ideology that presents nonwhite segments of the population as a dangerous underclass under suspicion and as a population undeserving of social benefits. However, this strategy was not immediately perceived as the racist system it was. How did the Dutch state manage to obfuscate its racist policies and pass them off as aseptic neutral practices that only sought to increase efficiency?

The Invisibilization of State Racism

The citizenship variable used by the Dutch government to determine the risk of benefit claimant fraud embodies in its dialectical simplicity the complex set of racializing technologies inherent to neoliberal racism. During the SyRI process, the Dutch government argued that the question on the national origin of claimants did not involve potential risks of discrimination, as the notion of citizenship is not related to those of race or ethnicity. This narrative is an example of what Eduardo Bonilla-Silva has described as color-blind racism.[55] Bonilla-Silva explains that color-blind racism conceals (and protects) racialized structures of oppression and the role they play in the socioeconomic order by maintaining an unequal social distribution that disproportionately benefits white people while punishing others. Color-blind racism operates by concealing the material and structural consequences of the unequal distribution of goods under a veil of abstract liberalism that frames identity production and racialization processes as the result of individual decisions and biases.[56] Rhetorically, color-blind racism distances itself from more explicit xenophobic narratives without questioning the material basis that sustains the racial structure. According to the color-blind ideology mostly adopted in Silicon Valley, racism would be overcome in a system that granted equal opportunities to all, for example, by making the ethnicity of each user invisible. Of course, this epistemology deliberately ignores the social and structural dimension of markers such as race and gender. It individualizes these identities, as if they exist in a social vacuum and are only the result of personal choices or cultural constructions. Color-blind racism also ignores how "neutral variables," such as citizenship or no citizenship, hide deep and dense historically rooted relations of inequality.

The Dutch yes/no variable, for example, ignores (or even erases) a whole history of racial and colonial oppression at the root of migration networks between the European Union and its former colonies as well as the political economy underpinning the heavily racialized labor market. Brett Neilson and Sandro Mezzadra explain how the unequal global distribution of labor and privilege on which contemporary capitalism depends operates through border processes producing a hierarchy of subjects—from the European citizen with the ability to work, receive benefits, and move freely to the economic migrant and refugee, deprived of these same privileges and always subject to surveillance, suspicion, and premature death.[57] Nationality is not, as the Dutch (and many other governments) claim, a neutral concept. The ways in which nationality has been shaped and is currently obtained are a source of heated political debate. While the European Union recognizes the free movement of people and capital among its member states, the same is not true for the former colonies, which until very recently were (and in some cases, still are) governed directly or indirectly from the metropole.

Despite being subjected to the political, social, and economic domination of the European metropolis, colonial subjects have been largely excluded from obtaining metropolitan privileges such as nationality and the more pragmatic civil, social, and economic rights that go with it. These subjects have been subjected to colonial domination, epistemic hegemony, and religious imposition; they were mobilized in wars and used as cheap labor but not granted the rights and freedoms enjoyed by their metropolitan counterparts. The exclusionary pathos of European nationality persists today in the form of jus sanguinis (prevalent in most EU countries) whereby citizenship is granted not on the basis of place of birth but on the basis of the nationality of the parents. Therefore, the use of nationality as an indicator to determine the risk of a subject to commit tax fraud is not simply a discriminatory act, it is evidence of a deep and structural phenomenon, an example of the racist habitus of the Dutch government. That is, it is one *more* element within a broad set of practices, behaviors, actions, and laws designed to maintain and validate a structure of dispossession, and to retain the social benefits of a cheap, racialized, and exploitable workforce. This is what differentiates the discriminatory outcome of an isolated case—what scholars have termed algorithmic bias—from the structural outcome of what some aptly refer to as algorithmic oppression.[58]

Austerity, Neoliberalism and Welfare Surveillance

There is a vast and extensive literature demonstrating how, historically, surveillance has played a crucial role in disciplinary institutions designed to extract value from, guard, maintain social reproduction of, and foster "social hygiene" among poor and racialized populations. The architecture and regulatory techniques of institutions such as the plantation, school, factory, hospital, or prison were especially designed to facilitate the surveillance of bodies and the imposition upon them of a regime of routines, control, behavior, work, and sleep. The control and surveillance of these dangerous, dirty, impure, or useless bodies extended beyond the walls of the institutions themselves through monitoring, reports, visits, espionage, denunciations, and tests, among many other strategies. The rise of neoliberalism in the 1970s represented a turning point in the governance of subjugated populations, extending neoliberalism's influence both in the United States and the rest of the Global North. In the US context, the Ronald Reagan administration played a prominent role in producing and popularizing neoliberal narratives such as the myth of the Welfare Queen. This fictional figure, depicting a single mother who allegedly squandered public funds on vices and a deviant lifestyle, was used to undermine the solid image of the welfare state and give way to the legitimizing narrative of significant cuts in social benefits.[59]

Similarly, in the United Kingdom, both the Labor Party and the Conservatives adopted discourses blaming those deemed undeserving for undermining traditional values associated with wage labor, whiteness, and patriarchy. These discourses not only influenced public perception but also provided ideological justification for making cuts in social assistance, setting efficiency targets, and adopting specific guidelines under the premise of efficient state management following the new theories of public policy management. However, these policies not only resulted in social cuts but also catalyzed control practices that intensified surveillance over welfare recipients. The Personal Responsibility and Work Opportunity Reconciliation Act (PRWORA) in the United States, for example, established stricter requirements and limited the time of receipt of aid, while technologies such as the Client Registry Information System-Enhanced (CRIS-E system) employed in Ohio facilitated extensive data collection on recipients, thus consolidating a surveillance regime and offering a veritable informational panopticon (and control) of, "among other things, veteran status; living situations; household income and expenditures; age, names and Social Security numbers of children; health information; work history; marital status; race; criminal history; divorce history; medical insurance; saving and checking accounts; burial contracts; cemetery lots; life insurance; Christmas clubs."[60] Simultaneously happening was the phenomenon of mass incarceration, that is, the radical increase of incarcerations as a consequence of deliberate criminalization and counterinsurgency policies, catalyzed in large part by the "war on drugs" and the privatization of prisons.[61]

The 2008 economic crisis revitalized the welfare state vigilante strategy implemented in the previous period. Austerity measures imposed by international and regional authorities generated an environment of severe budgetary constraints, unleashing market-oriented public policies and the privatization of public services. On the other hand, the rise of the war on terror and its punitive apparatus favored the development of mass surveillance technologies, multiplying their presence in all state structures, including schools, hospitals, social workers, taxes, and social services in general. This expansion of punitive surveillance gave rise to "digital welfare surveillance," where digital technologies are used to monitor and manage social welfare. This regime has been brilliantly described by Virginia Eubanks in her work "Automating Inequality," where she highlights how the implementation of digital technologies in the management of social programs has exacerbated the neoliberal regime of dispossession by amplifying the scope and granularity of surveillance and punishment techniques specifically designed to repress marginalized communities.[62] In short, the welfare surveillance regime was consolidated through policies designed to strengthen surveillance and punishment of the most vulnerable and marginalized sectors, contributing to the creation of a disciplinary system that regulated, controlled, and limited the lives of those produced as risk. In other words, neoliberalism shaped social policies in the service of surveillance and discipline of the poor while deregulating and giving carte blanche to the rich, a trait that today's digital capitalism has not hesitated to adopt as its banner.

Islamophobia

Since 2010, a multitude of far-right movements have been gaining power and influence, capturing the public debate on security, criminal justice, borders, welfare, education, and even public morality. The extreme right is not a homogeneous ideology; Bolsonarism in Brazil is not the same as the policies of Viktor Orban in Hungary or those of Javier Milei in Argentina. Despite the differences, there are a number of parameters and commonalities that make them recognizable. Keeping distances, the extreme right claims to be the last bastion of tradition and national essences and in this sense demands social cuts and an iron fist against the enemies of the country.[63] In Europe, these enemies are often identified with racialized people of Muslim religion, class organizations, sovereigntists, LGBTQIA+ people, or environmentalists. When political circumstances arising from neoliberal failure result in social protests, neoliberal and far-right governments and administrations often divert attention to alleged dangers lurking in the nation.

One of the best-known and most representative examples of neoliberal state Islamophobia has taken place in France throughout the twenty-first century. Although debates around the use of religious symbols in public spaces were not unknown in the secular France (see, for example, the controversies during the presidency of Jacques Chirac, 1995–2007),[64] it was during the Great Recession and its aftermath that the hijab-burqa controversy became a major national issue. Amid escalating protests against unemployment and precariousness and in defense of the countryside and the public sector, the government chose to attack the weakest. Regulations on religious symbols, previously ignored, became widely enforced and effectively banned in public spaces such as schools. These policies were subsequently upheld by the European Court of Human Rights.[65] The "Islamic veil" was presented as a threat to French and European identity. The polemic was constructed as a media product of utmost relevance, for which every self-respecting intellectual had to have an opinion. This manufactured identity crisis made invisible the very serious material consequences of the real crises experienced by the most vulnerable populations. Neoliberal governments read in an identity-based and nihilist key the increasingly frequent social protests in the banlieues, which were met with a huge wave of repression. France, like Spain, Belgium, the Netherlands, the United Kingdom, and Germany implemented a form of governmentality based on securitization, that is, the European iteration of the war on terror, consisting of the militarization of poor and racialized neighborhoods, the deployment of racist and classist discourses, and the dismantling of the welfare regime.[66]

Islamophobic discourses were used as a hook for the hundreds of thousands of people, who were witnessing firsthand the degradation of their living conditions, the soaring unemployment rates, and the unstoppable decline of the former industrial regions. In this context, neoliberal opportunism met Islamophobia, setting the stage for a new wave of heavily racialized austerity policies.[67] Geert Wilders, the charismatic leader of the Dutch far right, put it bluntly: "We could have been swimming in money and instead of doing so we follow the leftist's dream to get half the Islamic world to the Netherlands. The more voting cattle for the leftist church, the better. [. . .] But Chairwoman, who is paying the bill, who is paying that 100 billion? Those are the people who built up Holland, those are the people who work hard, the people who save up properly, who pay their taxes as they should, the common Dutchman who is not getting things for free: Henk and Ingrid are paying for Mohammed and Fatima."[68]

This discourse has progressively permeated the European electorate, moving from relative marginality to political centrality. Geert Wilders's party was the most voted in the Dutch elections of 2023. Table 1.1 offers a nonexhaustive list of far-right political organizations represented in national and European parliaments that have championed the Islamophobic cause. The list includes both established historical formations and new movements.

TABLE 1.1: Far-right political organizations with representatives in the National and/or European Parliament as of December 2023

Country	Party	2023—National Representatives	2023—European Representatives
Austria	Freedom Party of Austria	Yes	Yes
Belgium	Vlaams Belang (VB)	Yes	Yes
Bulgaria	Revival	Yes	No
Cyprus	National Popular Front	Yes	No
Czech Republic	Freedom and Direct Democracy	Yes	Yes
Denmark	Danish People's Party	Yes	Yes
Denmark	The New Right	Yes	No
Estonia	Conservative People's Party	Yes	Yes
Finland	Finns Party (Government Formation)	Yes	Yes
France	Rassemblement National	Yes	Yes
Germany	Alternative for Germany	Yes	Yes
Greece	Greek Solution	Yes	Yes
Hungary	Fidesz (Government Formation)	Yes	Yes
Italy	Lega Nord (Government Formation)	Yes	Yes
Italy	Brothers of Italy	Yes	Yes
Netherlands	Party of Freedom	Yes	Yes
Poland	Law and Justice	Yes	Yes
Portugal	Chega	Yes	No
Romania	Alliance for the Union of Romanians	Yes	No
Spain	Vox	Yes	Yes
Sweden	Sweden Democrats (Government Formation)	Yes	Yes

Despite its enormous heterogeneity, neoliberal racism nevertheless maintains numerous common features. First, despite its protectionist rhetoric, it is characterized by an intensified agenda of privatization, market dominance, and the prioritization of individualistic values over collective welfare. It also promotes punitivist approaches such as workfare and prisonfare to address poverty, which it articulates through technologies inherited from the colonial era conveniently adapted to the algorithmic moment. The expansion of neoliberal racism in the nations of the Global North is giving rise to a digitized and updated version of itself through a technologized collaboration between institutions, corporations, repressive state apparatuses, tax authorities, employment services, and other actors involved in the surveillance of the population constructed as dangerous. Instead of turning its energies toward establishing better mechanisms for redistributing wealth or prosecuting white-collar criminality, this strategy places subaltern and racialized populations in its crosshairs. Suppressing migration, controlling "fraud" in welfare systems, and maintaining "order" in the streets are the pillars of an increasingly digitized punitive approach disproportionately affecting impoverished and racialized populations.

Beyond Liberal Criticism

Contrary to the neoliberal narrative of efficiency and productivity, the digitization and automation of public services has been revealed as a costly and unsustainable system, socially, politically, and economically. This is something on which numerous academic as well as institutional voices agree; their critique is neither new nor marginal. Already in 2019, Philip Alston, UN special rapporteur on extreme poverty and human rights, drew attention to the threats to human rights arising from the advance of digital capitalism in the welfare state.[69] The report emphasized that decisions made by algorithms can result in unfair denials of essential services, disproportionately impacting the most vulnerable groups. Likewise, the rapporteur vehemently denounced how the lack of transparency and accountability and the asymmetry of knowledge and power in digitized welfare systems limits people's ability to challenge automated decisions that affect them. For the rapporteur, this undermines procedural justice by threatening (among others) the rights to due process and fair treatment.

In the academic field there is a whole new literature devoted to questioning algorithmic bias and lack of transparency in automated decision-making. The concept of "algorithmic fairness" has emerged as a mainstream topic in debates about ethics and fairness in automated decision-making systems. It is a recent but vast field, proof of which is the nearly five thousand related academic publications that appear in a simple Google Scholar search, published in the period of 2022–23 alone. Among the most prominent authors within this genre is the influential Frank Pasquale, who has highlighted how opacity in reputational, financial, and search algorithms dilutes the responsibility of states and corporations in the face of potentially biased machine decisions.[70] In the same vein, Cathy O'Neil demonstrates how socio-technical systems used in sectors such as finance or the criminal justice system can contribute to the perpetuation of regimes of inequality.[71] The notion of algorithmic fairness has been adopted by some of the great representatives of digital capitalism. This is the case of Microsoft, which has been defining the meaning of algorithmic fairness, transparency, and equality for years through the contributions of its FATE (Fairness, Accountability, Transparency, and Ethics) project. It is paradoxical to see how, on the one hand, Microsoft funds the "critical" research of researchers, such as Kate Crawford, who points out the social and environmental dangers of AI,[72] while at the same time, it collaborates with Israel on their technological war against the Palestinians and participates in fossil capitalism hand in hand with ExxonMobil.[73] Microsoft is certainly not the only company to have adopted the Algorithmic fairness flag. Google, Amazon, even the cyberwarfare company Palantir speak unabashedly of "Human-centered," "accountable," and "responsible" AI while closing contracts with the defense department and border agencies and deploying AI tools to exploit workers.[74] Various states and supranational institutions have also identified the need to take action against "algorithmic bias." For example, the Australian Human Rights Commission highlighted the importance of addressing algorithmic bias and establishing effective safeguards to protect human rights.[75] At the end of 2023, both the United States and the European Union (with its renamed AI act) published different legislative instruments aimed at moderating the most harmful effects of the increasing ubiquity of artificial intelligence by regulating these technologies.

Despite the undeniable relevance and pertinence of many of these arguments, the liberal critique of algorithmic domination falls short, limited in scope and depth. Its defenders see it only as a threat to individual freedoms reified by neoliberal dogma. The liberal critique fails to appreciate the structural dimensions of algorithmic domination as a form of class, race, and gender oppression. This formal recognition of a matrix of inequality affecting individual expressions (e.g., of identity) ignores instead the massive social harm perpetrated not as a systemic error but as the predictable and desired outcome of the system. The liberal critique fails to understand how the emergence of the digital welfare state is giving rise to the production of new forms of racialized exploitation and subjectification and is consequently incapable of taking the necessary measures to identify, stop, and repair the social damage caused.

Let's go back to the SyRI case. Beyond the spectacular resignation of Mark Rutte (who, by the way, was reelected shortly after his resignation) and his cabinet, the Dutch authorities imposed a fine of €2.75 million for discriminatory and illegal data processing on the Dutch Tax Administration.[76] The amount of €2.75 million was the limit of liability willing to be assumed by a wealthy European state, in what was one of the most blatant and obvious examples of institutionalized racism and classism. There was no in-depth review of the racist neoliberal policies that led to the emergence of the socio-technical system, nor was there any reflection on the deficiencies in the supposedly human supervision of automated decision-making systems, nor did it lead to a review of the widespread racist culture that, by the Dutch state's own admission, permeates its institutions. In fact, the Dutch foreign department as well as municipalities such as Rotterdam have continued to use SyRI-like systems to monitor the granting of grants and visas.[77]

The framework of abstract liberalism that pervades notions of algorithmic fairness or algorithmic justice is simply insufficient to challenge the regime of classism and structural racism that organizes socio-technical systems such as SyRI or Robodebt. We need new epistemic and political tools that, going beyond individualistic approaches, allow us to identify and name the socially harmful consequences shaking the lives of hundreds of thousands of people. This shift must also contribute to the demystification of technological solutionism and blind trust in automated correlation processes. We also need a new socio-legal understanding of the relationship between state and corporate crimes and the harm caused to their victims, both individual and collective. How do we deal with cases where it is the system—from grassroots officials to political leaders and the ideology that inspires them—that entirely fails, that is entirely responsible? How do we measure and quantify social suffering when it affects tens, hundreds, thousands of people as in the cases of Holland and Australia? What is the price of years of stress and nervousness? How much is the time spent on political organizing, on building demands and protests, worth? Is it enough to compensate people economically for the damage suffered? Or should we rather imagine new ways of dealing with the crimes of the powerful? The answer to these questions will not be found in criminological theory, sociology, or law, nor in corporations like Microsoft or institutions like the European Union. The answer lies in emancipatory practices, in transformative struggles. The answer to these questions will be found in the movements that are already demanding and building new ways of understanding the relationship between technology and society, between justice and democracy, between State and people.

2

The Digital Takeover of Education

The year 2019 was destined to be the end of technological innocence. The shadow of the Cambridge Analytica scandal was flying over the front pages of newspapers. It was no longer a suspicion or a conspiracy: corporations like Facebook spied on their users on an industrial scale, trafficked their data, and used their vast domain over online communications to profit. In the popular imagination, Silicon Valley was no longer the setting for the benevolent nerd-geniuses of The Big Bang Theory. Palo Alto was not the world headquarters of a new humanist philanthrocapitalism built on a utopian sharing economy as it had been promoted. The Bay Area had become the cradle of a new generation of billionaires who did not hesitate, and even bragged, about breaking any legislation that crossed their path; CEOs who lied before parliamentary committees of inquiry; and an industry at the service of the click, viralize, and profit generation, even if it meant facilitating the path to white supremacists, collaborating closely with global far-right campaigns such as Donald Trump's or Brexit, or turning a blind eye to governments that systematically violate the human rights of tens, hundreds, or thousands, such as Israel and India. In 2019, criticism of the then so-called GAFAM (Google, Apple, Facebook, Amazon, and Microsoft) went beyond the narrow academic and activist niche to become a major social concern. It was not just about awareness in the face of the influence of social networks or the growing devotion we all bring to our smartphones. Nor was it about high-profile legal proceedings such as those initiated by the EU against Google, Apple, or Facebook over competition issues. Digital capitalism became a material, palpable problem. Silicon Valley was on the streets of half the world, and not only on the screens.

In Barcelona and Madrid, neighborhood associations organized to protest against the gentrification of their neighborhoods, accelerated by the multiplication of tourist apartments promoted on AirBnb. In London, Mexico, Sao Paulo, and Paris, thousands of cab drivers mobilized against unfair competition and the lack of regulation of Uber. In Italy, Germany, and Spain, Amazon workers mobilized against precarious working conditions in warehouses and distribution centers. In 2019, in governments, unions, and civil society, there was growing discontent against the technological dominance of a handful of corporations and the growing role they had come to occupy in our lives. However, something unexpected happened.

The arrival of a global pandemic dislocated the scenario. In one fell swoop, old fears and bitter grudges took a back seat. Half the world went online, full-time, 24-7. Food delivery, long-distance deliveries, work, leisure, pleasure—almost everything became mediated by one private corporation or another. Where once there were scruples and concerns with private issues, with addiction, with corporate power over the infrastructures of the social fabric, there was now unquestioning surrender, total acceptance of the leonine conditions of digital capitalism. This is something you surely experienced; I certainly did. We know that Facebook traffics in data and watches over us in its apps, and that Google does the same every time we enter its inevitable overlapping constellation of the web. But what else could we do from our individual positions—live disconnected in our enclosures? Impossible; for many it was not only a personal decision, but a work obligation. Moreover, immersion in the corporate universe was not something new but the expanded continuation of dependencies (on WhatsApp, Netflix, Gmail, etc.) that had long been embedded in our daily lives. Yes, it can be argued that as individuals in the digital society we are users and consumers rather than citizens. Such is the power of digital capitalism to colonize, to take, to hegemonize. An important question immediately arises. Does this sort of historical inertia that seems to push us as individuals to corporate dependence also subject us in the collective dimension? Has it also managed to bend the institutions? For many political leaders the answer is yes. Think of education; during the quarantines, the forced closure of educational institutions forced education providers at all levels to adapt quickly to a distance learning environment. What can we do? How do we proceed? The response of public authorities in countries as diverse as Australia, Spain and Argentina was similar. Beyond providing a few programmatic guidelines, rules, and a handful of tips, the truth is that in practice, educational centers were left to their own devices.

Collective shock coupled with public neglect created fertile ground for corporate opportunists to step in and fill the state's void. Digital capitalism made a philanthropic offer that society could not refuse: a digital educational infrastructure ready to be used, and best of all . . . for free! In a lightning-fast manner, criminal corporations like Google, already facing billion-dollar fines by then, took over the educational infrastructure of more than 150 million students. In the new platformized state, democratic rules were replaced by proprietary code, and educational inspections by surveillance capitalism.[1] Was this really the only possible option? To entrust the means by which the education of hundreds of millions of children would take place to corporations that had already been convicted of having, among many other things, exploited children's data for commercial purposes? Was this really inevitable?

Despite justifications, this was not the result of chaos, circumstance, or haste. All technology, all infrastructure, all science, is politics by other means. If there was not total and free Internet for all, if there was not a communication and entertainment network that did not demand economic or data payment, if there were no communication tools guaranteed as a universal service, it was not for lack of capacity, technologies, or infrastructural power. It was the result of the surrender, the *total* surrender, of the political leadership of a large part of the world to the private logic of digital capitalism and its techno-solutionist mantra.[2] Evidently, the commodification of the Internet did not happen in two days. Its most immediate antecedents are to be found in the privatization processes that led in the mid-1990s to what had been a global experiment financed largely by public funds becoming a market good, something to be exploited. As important as it is to know about this aspect of privatization and plundering of the public, which generally remains unknown, this is not what I am going to focus on. (If you are interested in going deeper into this subject, I recommend reading *Internet for the People: The Fight for Our Digital Future* by Ben Tarnoff.[3]) What I am going to talk about in this chapter is how, growing on the foundations of privatization of technological infrastructures and taking advantage of a context of crisis, digital capitalism is expanding its influence over the public sector. Specifically, I will analyze how the logics and architectures of Silicon Valley digital capitalism are colonizing the educational system. I will start at the end, with the harmful effects that digital capitalism is imprinting on workers in the education sector.

Autoethnography of the Digitized Educator

Between 2016 and 2021, I had the opportunity to work for tertiary institutions in the design, delivery and assessment of online content. This experience allowed me to witness firsthand some of the so-called "advantages" of flexible, digital academic work. One of the most notable features of this arrangement was the absence of an actual employment contract. As in many other cases of platformized work, online academic service providers do not have large workforces. Beyond marketing and some management positions, most of the actual work is outsourced. The strategy of service providers is to fragment educational tasks into operations of varying responsibility, duration, and size, such as teaching one or more modules, tutoring students, directing theses, moderating forums, designing or coordinating courses, and so on. These tasks are not handled by a small but collegial body of permanent professional workers, but by a myriad of academic gig workers considered as freelance providers. That was my case.

It should be noted that these "contractors" and "adjuncts" constituted the bulk of the workforce; in other words, they were totally indispensable for the course to come to fruition. In my case, despite being considered a totally external and autonomous worker I was responsible for about 20 percent of the student credits for the semester. Another intriguing aspect of my contract, shared with other forms of platformed work, was that I was paid not for my time, but for each piece of work completed, for example, each accepted thesis that I supervised. As research into the platform market has recently shown, this form of wage structure benefits capital intensification while relieving the capitalist of any responsibility to the worker.[4] To this, it must be added that under the telecommuting formula, the capitalist externalizes costs necessary for the work—energy, internet connection, heating—to the worker. Savings are not limited to this. The capitalist also remains comfortably alienated from the whole range of social rights associated with the figure of the employee. Basic conquests of the workers' struggle, such as sick days, vacations, and maternity or paternity leave, are out of the question in the new digitalized precariousness.

The curious thing is that, despite the degradation of working conditions, somehow the academic value of my training was still recognized, even if only to exploit it. For example, I was offered on different occasions to design and write the content of online courses. The task of curriculum design in academic institutions normally falls on senior faculty; in some cases, this responsibility is left to early-career researchers, but in no case to contingent workers. I was initially attracted to the idea, but then I carefully read the fine print. It wasn't just a meager payment; the worst part was that I was obliged to assign them the intellectual rights to my work without receiving any kind of recognition. The corporation would reap the benefits of my creation without even mentioning my name. Meanwhile, other precarious educators would teach the classes. In other words, fracture and division worked as a capitalist's strategy for exploitation.

Isolation is one of the fundamental mental traits of online teaching; it means weakness for the workforce and power for the corporation. Theoretically, teaching and assessment activities are coordinated by a senior academic staff member, but the reality was otherwise. Adjuncts supposedly providing a small fraction of the tasks were actually charged with fulfilling their responsibilities themselves, with only a vague, and often nominal, connection to an administrative person detached from the academic plane with whom they communicated remotely. The opacity and information asymmetry are total. The professors who are platformed are unaware of the structure and functioning of the institution for which they work. The management company handles all the economic and academic information. They control the data, the contents, the salary we receive as teachers, and the prices the students pay. Information for the rest is inaccessible and opaque—we do not even know who the educators assigning the other parts of the grade are.

Aware of the fraudulent reality of their business model based on the precariousness of the teaching staff, companies use a myriad of strategies to avoid labor regulation. For example, adjunct faculty are often renamed under other formulas such as facilitators or mentors, a practice that seeks to minimize their role. These formulas are a mere artifice, and in fact they are not used by students, who always address the supposed facilitator as professors in emails and forums. In spite of playing a fundamental role in the teaching and evaluation of the courses, the *facilitators* and *mentors* do not have institutional visibility on the websites of the universities where the academic heads of the degrees do appear. The names of those who will be the actual faculty are provided to the students exclusively through internal means, in the form of welcome messages on the platform or emails. This strategy serves a double function. First, it makes invisible the academic work that is fundamental to the achievement of the course. Secondly, in the absence of physical spaces, it prevents workers from contacting each other.

Despite these hurdles, I informally contacted several adjuncts only to discover that they were, like me, nothing more than a precarious and disposable digital workforce. For several semesters, I was entrusted with mentoring and evaluating students' final assignments, which in practice functioned as a master's thesis. Paid by the piece, adjuncts were only rewarded if the student completed, submitted, and, if necessary, resubmitted their final assignments. This logic is totally perverse. Paying per piece based on student success encourages adjuncts to pass anything that comes their way, regardless of quality. After all, if a student fails to achieve the minimum grade, he or she is entitled to resubmit, which means unpaid work for the evaluators.

While perhaps less precarious, the most renowned institutions also follow practices that expand the precarity of a growing irregular academic workforce. As several unions have pointed out, the inclusion of data-extractive platforms such as Zoom, Turnitin, and Google as part of everyday teaching has resulted in workers being sucked into the vacuum of surveillance capitalism.[5] Now, teachers must create accounts, learn to use platforms, create content, and manage online teaching while private platforms spy on them in endless ways. This forced adoption of digital technologies has not only taken a toll on privacy rights but also on mental health, with Internet-related stress linked to teaching and research activities. The logic of surveillance also penetrates the internal rationale of teaching and research working conditions. The datafied student, whose data and metadata are visible and accessible to faculty and adjuncts, is just one small example within the hypertechnologized panopticon in which university personnel are as much prisoners as guardians. And in the tower, hidden from view, are the corporate university managers and technology giants.[6]

As a teacher, I had a granular view of each student's login, their last access, the duration of this access, their participation in forums, and so on. What information was collected about my work time and by whom?

How is this information used or how will it be used in the future? Neither unions nor workers have much of a clue about these processes beyond the techno-solutionist rhetoric that the greater the concentration of data, the better. I for one am suspicious. The history of the workers' struggle in the last three hundred years can be read from an infinite number of angles. One of them is the struggle of the capitalists to have more and more inquisitive, meticulous, and detailed control of the labor force. More vigilance, more control, more productivity, less resistance. This has been one of the many equations of industrial managerialism: it is always prone to the panopticon, to surveillance, and to domination, even if this is exercised in unexpected places.

One of the areas where this zeal for quantification, surveillance, and corporate managerial record keeping is best revealed occurs with the researchers themselves.[7] Today, almost anyone involved in academia and born after 1980 is expected to have a Google Scholar profile. Circumventing this new obligation is not only impractical in the face of applying for grants and jobs, but it can generate suspicion, suggesting laziness, lack of productivity, or even ignorance. I think it is easily argued today that research is the first aspect of academic work to be completely subsumed in digital capitalism. Under the regime of numerical governmentality, scientific research is treated as a popularity contest. Researchers are rewarded according to the number of citations their work has obtained (and have been captured) by platforms such as Google Scholar. Their work, their trajectory, is reduced to an index algorithmically generated by proprietary rating systems.[8] An opaque, unquestionable system that has nevertheless become the currency of academic success. But not just any publication is enough, not just any citation. Only those publications with a high impact index will have value. As in the case of research bodies, these ratings are produced artificially and algorithmically by corporations dedicated to data processing analysis. These processes take into account factors such as the reputation of the journal when awarding their numerical rating.

That is, academic success is not defined by contribution to a field of knowledge but is quantified by digital platforms using aggregate metrics that measure the number of published articles, their citations, and the perceived relevance of the journals in which they are published.[9] The most commonly used metric, the h-index, is defined as a "The h-index is a measure of the number of publications published (productivity), as well as how often they are cited. h-index = the number of publications with a citation number greater than or equal to h. For example, 15 publications cited 15 times or more, is a h-index of 15."[10]

Trying to understand the algorithmic methodology used by these knowledge brokers to quantify the productivity of academics at the global level represents a considerable challenge. The issue becomes even more complicated if we try to define the elastic concept of "impact." What is the true meaning of this elusive word, valued equally by the academic rockstars of the TED talk scene and the dour European Commission? What is more impactful, changing people's lives or being read by a select group of influential gentlemen in prestigious institutions? Today, even academics who maintain a critical stance are compelled to conform to standards such as the h-index, directing their research toward a limited number of journals dominated by an oligopoly of publishers who profit from publicly funded research. In other words, academics are pushed to privatize knowledge, often financed by public funds, under the threat of not getting a decent job. An extractivist process that feeds on both the work of researchers and public goods.

Sociologist Robert Ovetz offers a critical perspective on these phenomena by situating them within broader transformations in the technical composition of labor.[11] Following Karl Marx and Harry Braverman, Ovetz argues that today's technological transformations share similarities with the capitalist aims of the nineteenth and twentieth centuries, which sought to reorganize industrial processes while undermining workers' organization and resistance to domination. These transformations aimed to break down artisanal manufacturing and workers' expertise into fragmented tasks that could be easily performed by new machines, with the assistance of unskilled and replaceable operatives; this is what we know as the proletariat. The online education sector appears as a field of experimentation where central skilled tasks of academic work such as designing, delivering, and evaluating content are chopped up and packaged, ready for delivery to precarious, replaceable figures on demand. The fragmentation of teaching tasks serves the dual purpose of cutting costs and alienating academics from their work, tearing apart what should be organically connected elements of the course. The parallel between nineteenth-century proletarianization and that which is taking place under digital capitalism is more than evident. A handful of capitalists have put in place mechanisms to intensify exploitation and multiply profits while eroding and dispossessing the working class.

Ovetz also pays particular attention to how learning management system (LMS) technologies are designed to "rationalize academic labor by transforming the assessment of comprehension of content knowledge to measurement of proficiency in task completion."[12] According to Ovetz, this pedagogical transformation seeks to produce more productive and self-disciplined learners molded to meet the growing demand for platformized work.[13] That is, a change at the teaching level is being deployed to produce the digital proletariat of tomorrow. Of course, it is difficult to generalize from personal experience, even if it is accompanied by a growing number of jobs such as those already discussed. Nonetheless, these technological and labor transformations in education prompt an examination of how the seemingly innocuous new teaching opportunities under digital capitalism hide profound changes that affect the very nature of educational work. In the next section, I will zoom out again to trace a brief history of the political economy that is enabling digital capitalism's assault on the public sector and, more specifically, the educational sector.

Political Economy Creeps beyond the Line

In 2009, journalist Jeremy Caplan published a terse article in the *Times* titled "Google and Microsoft: The Battle Over College E-Mail."[14] The article was not a big hit, there were no big statements, nor was there (apparently) a big scandal uncovered. Google was not yet the giant it is today, and the *college e-mail* issue seemed more of a niche topic than indicative of a vast socio-technical transformation. But perhaps for that very reason, the article is profoundly interesting on many levels, and I would even go so far as to say that it is fundamental to understanding the early expansionism of digital capitalism in the public sector. According to Caplan's surprisingly fluent account, Google, a company barely five years old, was already managing university mail for more than two thousand US universities and colleges, including some of the most prestigious names. More surprisingly, 42 percent of educational institutions had already outsourced this service to one digital corporation or another. At first glance, it looked like a win-win situation. Emails were becoming a fundamental tool in digitized educational environments by leaps and bounds. It wasn't just about communications; papers, essays, and projects were increasingly taking place in these environments. University servers and applications simply couldn't cope. Students and staff were ditching cumbersome internal email services in favor of a learning environment based on the powerful search engines of the new platform capitalism. It wasn't just speed. Big tech was multiplying the limited capacity of university servers. The transition to Microsoft and Google seemed not only inevitable, but desirable. This information is already valuable in itself, but the most interesting part of the article is not in the narrative of the spectacular corporate expansion but in the story of the causes that make it possible: "Google's Apps for Education program has gained significant momentum as student tech demands mount and budgetary pressures strain campus IT departments. Handing the e-mail keys over to Google helps schools avoid costly server upgrades while capitalizing on Web-based e-mail's popularity among students."[15]

The article naturalizes, and presents as happy and inevitable, the result of broader macroeconomic policies. It is not, as it might first appear, about the battle of two corporations for "university e-mails." The central theme of the text is the massive privatization of university digital infrastructures in a context of cutbacks and outsourcing sponsored by neoliberal policies. This is the breeding ground of digital capitalism. The concept of free services resonated powerfully with the managerial class in charge of guiding the business-universities of the twenty-first century. Not by chance, the joyous revolution of email privatization coincided with the emergence and expansion of Google Books, Google Scholar, and soon after, Google Drive. After all, who can resist free? Apparently, at the time, no one.

Amid job cuts and the elimination of entire arts and social science programs, unions seemed to have more pressing problems to address. Besides, what arguments could they make for resisting the free provision of services, no matter how corporate the actor providing them? None, even less so if these actors maintained an aura of academic and even progressive respectability.[16] In fact, despite the obviousness of the privatization movement, the professional bodies of universities embraced the corporate landing with some enthusiasm. For example, the IT community at Stony Brook University uncritically copied Google's corporate advertising with a resounding: "Holy Cow! Unlimited Storage Coming to Google Drive."[17] Years later, when infrastructure delivery was already total, Stony Brook's website unabashedly collected the "Reasons to Love Google Drive"—but perhaps a forced love, a love against which there is no alternative.

Just a few years after launching its services, Google was already facing accusations from students at numerous US universities for allegedly violating their communications. It was alleged that the Google system systematically processed, scanned, and indexed mail, about which students, families, and administrators were not properly warned or informed. According to the plaintiffs, this invasion of their communications violated their privacy in breach of the US Family Educational Rights and Privacy Act. Google was not only unapologetic, but pointed out that this intensive processing of email content was a fundamental part of the business model and simply could not be turned off.[18] In fact, just a few months earlier the corporation made it clear people could not expect strict confidentiality when using its services. It was in the heat of that debate that Eric Schmidt, then CEO of Google (and now heavily involved in AI for military purposes), famously pronounced that the "Google policy is to get right up to the creepy line and not cross it."[19] Some of the media allied to digital capitalism, such as the *Verge*, came to Google's defense, calling for restraint and, yes, inviting the vanguard company of digital capitalism to do better.[20] In any case, these scandals began to highlight the consequences of handing over critical educational infrastructures to unscrupulous corporate entities.

Months later, and in the heat of the Spying on Students campaign led by the Electronic Frontier Foundation, it came to light that in addition to universities and colleges, these practices were extended to tens of thousands of elementary schools and high schools.[21] And, even more, Google was using all kinds of applications of its educational matrix to track and monitor the online behavior of minors (something I will delve into in the following sections). Again, Google's response was elusive, defending that its actions were within the framework of legality.

Despite the notoriety of these cases, school administrators continued to trust Google, not so much in their good faith and in their (false) statements presenting themselves as legitimate data custodians, but as

Figure 2.1: Schrom 2014.

providers of free unlimited storage on Drive.[22] What value do privacy rights have in the face of the promise of free digital infrastructures? Only three years later, in 2017, mainstream global media were already talking about a now self-evident concept, the Googlization of education. In a much-cited newspaper article entitled "How Google Took Over the Classroom," Natasha Singer described the corporation's new position as a dominant player in the US educational technology market.[23] What only a few years earlier had been a promising niche business opportunity ended up becoming the engine of the new educational technology (EdTech) field. It was a titan used by tens of millions of people, the vast majority of them minors. In 2023, Google provides not only email, but much of the digital infrastructure to approximately 150 million people worldwide.[24] Never before has a corporation (with the exception perhaps of the Jesuits in the Spanish imperial colonial world) had such an influence on global education.

Once settled in its position of hegemony, Google revealed its cards, that is, the price of free, the cost of dependence. Google unilaterally announced new conditions for its "free" services in universities. In a carefully crafted post on its corporate blog in 2021, Google signaled the end of free storage for educational institutions, something that would take place in July 2022.[25] The effects were immediate. Academic institutions around the world were quick to send threatening emails to their alumni and former employees warning them of the end of the "free" era and, with it, the end of access to services that only a few years earlier Google had guaranteed as perpetual. The University of Auckland, for example, informed its former staff, "We will block access to your files from 31 December 2022. We will then automatically delete your files 3 months after blocking your access."

Other more powerful and influential institutions, such as New York University, managed to negotiate moratoriums and possibly a preferential contractual framework.[26] However, most institutions addicted to and dependent on Google infrastructure were forced to choose between accepting worse and limited services and succumbing to Google's blackmail and paying for the services. How did this situation come about?

The Path of Servitude

In order to understand the recent privatization of educational infrastructures perpetrated by digital capitalism, it is necessary to situate the phenomenon within the broader processes of commodification of the public sector unleashed during the neoliberal wave of the 1970s. The neoliberal revolution was not content with altering the regime of priorities of the economic-political agenda, proposing, for example, extensive privatizations. Populist politicians such as Margaret Thatcher and in the UK and John Howard in Australia or dictators such as Augusto Pinochet and in Chile and Jorge Rafael Videla in Argentina understood neoliberalism as a profound transformation of values, morals, and the relationship between state capital and society. As explained by scholars of neoliberalism such as Gregoire Chamayou, Veronica Gago, and Michel Foucault, the market for this ideology, not only a mechanism of exchange of goods and services but the fundamental source of truth, the principle that guarantees the existence of the ultimate value, is freedom (or a certain idea of it).[27] This explains why under this dogma, market values are systematically placed above social values, which are considered, if not a burden, the chain of subjection, tyranny, poverty, and mediocrity. The neoliberal transformation not only aspires to the private management of all conceivable goods and services, material and immaterial, individual and collective—from water to health to forests—but also seeks to install its principles in the last redoubts of the public sphere. In other words, any organization must operate according to market principles.

Under neoliberalism, education ceased to be conceived as a technology intended for the traditional purposes of liberal democracies—such as contributing to collective welfare, reinforcing the values of the system, and promoting equality and social ascent. Instead, it was reimagined as a product to be governed by the same criteria of efficiency and monetization as any other commodity.[28] Universities and research institutions were portrayed as wasteful entities in need of close scrutiny, audits, and new forms of management in the name of "economy." Pressure for accountability and efficiency led to the adoption of corporate management styles, where performance metrics and results-driven strategies became the norm. The impact of this change was vast and profound. It affected financing models, reducing long-term public investment and encouraging private investment guided by immediate market applicability. It altered institutional structures, both in management and in the processes of internal democracy. Universities became populated by a new breed of managers and administrators more concerned with impact, ranking, and metrics than with the production and transmission of knowledge. At the same time that this managerial body grew, the body of permanent professors paradoxically decreased, initiating the long road to the precariousness and casualization that we suffer today. Even the rectors mutated from being renowned academics to becoming CEOs of corporate brands with millionaire salaries.

This intense neoliberal transformation of logics and principles in education also altered the conditions of the social pact between state and society. If education was just another commodity, why should the state provide it for free? In the 1960s, the neoliberal guru Milton Friedman, perhaps the first renowned intellectual to advocate transferring the cost of education to the students, must have asked himself the same question.[29] Friedman was rowing against the tide at a historical moment when, after decades of struggle and social organization, access to higher education was finally becoming widespread thanks to the reduction and even abolition of access fees. Nevertheless, Friedman's then extravagant ideas resonated strongly with the intrepid managers of the new policy, who in barely twenty years managed to reverse the hard-won social gains.

This is the case in Australia. The country, led by the Labor Party, experimented in the 1970s with one of the most extensive welfare states of the time. In 1974, and after arduous and massive interunion and student organizing, tuition fees were abolished in order to facilitate access to quality education for working-class and Indigenous youth. But the neoliberal virus was already installed in the system. In 1988, a system of "symbolic" fees was already introduced to indicate to the student body the monetary value of their education.[30] Beyond the practical consequences, this gesture was intended to prepare the transformation of education from a state principle to a commodity. Today, a law student at the University of Melbourne will accumulate a debt of approximately A$100,000 in fees alone. Similar movements took place in New Zealand (1992), the UK (1998), and Canada (1998) among many other examples.[31] Despite what it might seem, the commodification of education has not been a peaceful phenomenon. The measures were imposed with political and police violence and were always met with fierce resistance from unions and students. Notable examples include the epic struggles of 2010 in the UK[32] and the massive protests in Quebec in 2012.[33]

The neoliberal tsunami of the 1990s was not limited to tertiary education; its commoditizing waves reached countless public goods and services. From public enterprises in critical sectors such as energy, water, and transport, to the cultural sector—libraries and museums—to the sensitive areas of health and care. Public authorities were not alone. They made the journey towards the utopia of the market accompanied by the biased (and expensive) advice of large consulting firms and the influential voice of the corporate lobby.[34] The neoliberal era opened the ban on public-private partnerships, a crude conceptual allegory that could barely conceal the massive transfer of public funds to a handful of powerful corporate actors. All this was done in defense of the economy, efficiency, and the common interest, which for the evangelists of neoliberalism was equivalent to the private sector. Despite the promises of delusional, stellar futures that politicians like Fernando de la Rua, Carlos Salinas de Gortari, and Tony Blair made to justify public plundering, countless studies point to the devastating consequences of such decisions. These range from minor inconveniences such as those linked to decreased routes and frequency in public transport to the more tragic and irreparable ones derived from cuts in public health and care.[35] Despite the rhetoric and narrative, neoliberalism did not intend to repair the public, it desired its annihilation. It's useful to compare two headlines separated by fifteen years.

The first headline, from 2006, announces Tony Blair's welcome for private enterprise in the National Health System as the only remedy to repair something that was perceived as broken and on the brink of disaster. Even then, the aspiration was for private investment to account for 40 percent of total operations: "For us, the sterile debate between public and private health provision is over."[36] In 2022, another headline, in a much darker context, proposed, "Why Is Private Health Booming and the NHS in Crisis? Because That's What Ministers Want."[37] In that time frame, the degradation of public health services in territories such as the United Kingdom, Spain, and the United States is more than evident. According to a recent report referring to the United States, public control of hospitals decreased by 42 percent in the period 1983–2019. The most immediate effects of this privatization were suffered by lower-income or vulnerable patients, many of whom were no longer admitted because they were not profitable under market criteria. The second most affected were the medical professionals. Those who were not laid off saw their contractual conditions worsen and had to face casualization and downsizing.[38]

Primary and secondary education did not escape the neoliberal wave. The infiltration of this ideology managed to alter both the internal logics of classroom management and the educational model as a whole. Behind closed doors, the pedagogical model underwent a reconversion toward an economistic evaluation model of education based on the quantification of results and the permanent accountability of teachers.[39] An example of this is the proliferation of international, national, and regional standardized tests such as the Programme for International Student Assessment (PISA) and the Australian National Assessment Program – Literacy and Numeracy (NAPLAN) that supposedly evaluate the quality of education by reducing the learning trajectories of minors to a statistical grammar of quantifiable data.[40] This type of test not only serves to *measure* the quality of students but also to evaluate the quality of teachers, thus imposing a perverse logic that pushes teachers to a pedagogical design focused on passing tests. It is paradoxical to see how the neoliberal state, in the name and defense of freedom and "through the use of standards, assessments, and accountability, aims to restrict educators to particular kinds of thinking, thinking that conceptualizes education in terms of producing individuals who are economically productive."[41]

The overlap between neoliberalism and racial capitalism is also evident in this context, as critics of initiatives such as the "No Child Left Behind Act" passed during the George W. Bush era (2001–9) have shown. Following the parameters of neoliberal governmentality, this law sought to "improve" education in racialized and vulnerable sectors of the population by establishing specific evaluation mechanisms for centers with a special presence of this population. The legal instrument established a twisted system of rewards and punishments based on the fulfillment of teaching and student objectives set by the government. Those centers that did not meet the objectives would see their budgets reduced. In addition, the law allowed parents of students in school districts that did not meet the targets to transfer their children to other schools. As has been noted in countless papers, the predictable results of this initiative were devastating.[42] Many of the schools with the greatest challenges (and needs) were unable to meet the objectives and their budgets were reduced, thus worsening conditions for teachers and students. The law allowed the transfer of students from the most depressed centers to others as long as the former did not meet certain objectives. This facilitated the exodus of students from families that could afford it to prestigious schools, thus generating ghettoization and segregation.[43] The state thus shirked its responsibility for the adequate financing of schools by making students and teachers from poorer neighborhoods responsible for school failure.

The matrix of quantification and permanent evaluation is at the heart of current digitalized logics. The neoliberal capillarization of education was accompanied by less subtle privatizing and commoditizing processes. Examples of this privitization include the rise of charter schools in Republican jurisdictions in the United States,[44] the unstoppable increase of private schools in Latin America, and the transfer of public funds to charter schools in Spain, where in regions such as Madrid these schools represent more than 40 percent of the supply.[45]

The profuse privatization of public services was accompanied by intense outsourcing of infrastructure and professional expertise, particularly in the field of information technology (IT). State and institutional developers of the 1960s and 1970s became passive recipients of corporate technologies in the 1980s and 1990s, resulting in complete dependence on proprietary software, hardware, and expertise by the 2000s. To be fair, more than few administrations tried to weather the capitalist storm, seen, for example, in the early initiatives to develop tools and operating systems based on free and open-source software, such as Guadalinex in the Andalusia region of Spain. Despite their intrepidity, they were quickly eclipsed by the predatory and expansionist dynamics of technological giants such as Microsoft, which had declared this type of collaborative systems a cancer to be eradicated. This retreat from the public facilitated the replacement of public services with market-driven alternatives, paving the way for the current dominance of big tech.

If this scenario was already a neoliberal dystopia, the nascent realm of digital infrastructure looks even worse. Between 2010 and 2020, the percentage of college students who obtained comprehensively in online programs increased from 6.5 percent to 44.2 percent.[46] During the same period, the EdTech market skyrocketed from $500 million in venture fund investment to $16.1 billion.[47] Neoliberal universities, eager to expand their market, identified in EdTech platforms the technological and financial solutions to structural problems. Of course, it was a round business for neoliberal universities, corporations, and investment funds alike. On the one hand, universities were able to reach a wider audience, thus expanding and diversifying their revenue streams, all with virtually no costs (to them—costs were largely passed on to students or employees), something that would have been impossible at that scale given the disinvestment in endogenous knowledge and the outsourcing of digital infrastructures.⁴⁸ On the other hand, this has served as an opportunity for investment funds to increase both their market lines and to inject the logics of the financialized economy into the heart of the public sector.⁴⁹ For the corporations of digital capitalism, this has meant the possibility of capturing many of the dimensions of the education sector. It seems impossible to conceive of a university in the Global North or South (with the exception of some Asian countries) that does not make extensive use of proprietary software and hardware throughout its administrative and management processes, both internal and public-facing. Yet, digital capitalist corporations are increasingly present in digital teaching infrastructures (with an estimated reach of 80 percent of US institutions) and in the provision of online teaching services (with EdTech presence estimated at 80 percent).⁵⁰ This short-sighted strategy cements a more than predictable and unsustainable infrastructural dependence—a subjection that is at once economic, scientific, and social. First, it implies a significant transfer of wealth from the public to the private sector, a sunk investment that cannot be recovered. Second, it restricts the possibilities for endogenous technological development by reaffirming the dominant position of a handful of hypertrophied corporations. Finally, corporate software chains not only imprison the public sector to the logic of the market but also limit the horizons and experiences needed to face the challenges of the future.

These hypotheses were confirmed when the first COVID-19 quarantines and enclosures took place in 2020, when millions of educators, researchers, and teaching professionals were forced to move their activities to digitalized environments hegemonized by the private. It was then that the world realized that education depended on the educational infrastructure provided by the digital capitalism of Google, Zoom, Microsoft, Turnitin, and Coursera.⁵¹ Infrastructural dominance not only gives the private sector power to determine conditions and prices, it allows it to reimagine the role of the educator in new spheres, reconstructing them in its image as content creators, moderators, and coaches and, if not, as expendable and replaceable ghost gig workers.⁵²

The pandemic catalyzed, accelerated, and multiplied the process of digitization and privatization of education to levels that few would have predicted just five years ago. For example, online education, challenged from a myriad of fronts, has not only gained widespread acceptance but has also become a significant source of revenue for the education industry. What was once considered a secondary and despised option has now become the preferred choice for many students. Today, it is inconceivable for a college student, even in face-to-face settings, to lack immediate access to materials, recordings, and online work options with peers, tutors, and faculty. The digitized classroom, however, is not an innocent and neutral space. It is a classroom shaped by capital. Tablets, laptops, and learning management systems are written in a code whose sole stated purpose is to generate profits for their owners. Now, I will explore how the transfer of responsibilities and digital infrastructures from the public sector to private entities adversely impacts educational environments, affecting both those who work and those who learn there.

Googlification, Naturalization, and Families

The Australian Capital Territory government governs a territory with a population of nearly 430,000, which includes approximately 46,193 students enrolled in the public school system (in 2023). Under the slogan "Better Schools for Our Children," the progressive administration that governs the city has been pushing an ambitious campaign since 2016 called the Future of Education—Digital Access and Equity Program. The stated purpose of the campaign is to "continue to engage our high school students, enhance their learning experiences, provide them with opportunities to collaborate and discover, and in doing so produce responsible, literate and knowledgeable digital citizens."[53] This laudable goal was conceived with a specific objective: to ensure that every high school student in Canberra had access to a computer with an Internet connection. To achieve this, Canberra authorities partnered with Google, committing to provide Chromebooks to every high school student in the public network.

Why this device and not another? The justification for this undivided trust in a single provider to enable the instruction of the digital citizens of the future was not based on well-thought-out pedagogical opinions, nor on the calm democratic decision of parents and educators. It responded exclusively to the deals concocted between the public administration and the technological giant. In fact, the "parent booklet" offered to families by the Australian Capital Territory government did not offer an argument developed by any of the many experts and educationalists residing in the city but regurgitated arguments taken from Google's corporate publications.[54] That fact sheet highlighted in glowing terms the great performance of Chromebooks and the supposedly free nature of Google services for education. This is certainly not unique. It is easy to find similar information in institutional documents from many other jurisdictions. Canberra, in this regard, is just one example among thousands of territories and cities around the world that have opted for Google as a partner in managing their digital and blended educational environments.

For example, Burnside Primary School in Adelaide, South Australia, follows a BYOD (bring your own device—Chromebooks specifically) policy. Its website claims that "Chromebooks are increasingly being incorporated into schools because they increase students' anytime, anywhere learning capabilities."[55] As this poorly disguised advertisement boasts, online access in classrooms significantly broadens the scope of learning experiences, enabling creative thinking, communication, and collaboration. The Warwick Valley Central School District in New York (US) makes this unabashed case: "Chromebooks provide 21st Century technology skills to students through the lens of a safe technological environment inside and outside of school. . . . Chromebooks enhance classroom lessons and add to students' interest and intrigue in a variety of topics across curricula. . . . The Chromebooks and Google Apps open new and exciting ways for students to learn, preparing students for higher education and today's workforce."[56]

The celebration and uncritical acceptance of devices manufactured by a criminal corporation is deeply disturbing. However, the real cause for alarm lies in what lies beneath the surface. Chromebooks, replete with a proprietary ecosystem and administration, operate as Trojan horses, allowing Google to confuse itself with the educational environment, in effect making itself indispensable to it. In this way, Google not only sells devices and the services necessary to maintain them (often subsidized with taxpayer dollars) but also infiltrates the classroom and exerts its influence over those within it. Despite the significant contextual and structural differences between the various regions where education is being influenced by Google, several common problems can be identified.

Naturalization of Digital Takeover

Chromebooks, and, more broadly speaking, the digital invasion of classrooms, respond to the artificial—rather than pedagogical—need to digitize education. This is not neutral; it is a political decision that has been constructed through a narrative that advocates—as, for example, Australia's F-10 skills curriculum does—for helping students "participate in a knowledge-based economy and empower themselves within a technologically sophisticated society now and in the future."[57] This end that could be achieved in a myriad of ways, for example, with less invasive measures such as do-it-yourself workshops, is transformed under the logic of digital capitalism into the need to mediatize classrooms through digital devices. For example, the digital policy of a public school located in Melbourne's affluent inner north states that "use of digital technologies, including the internet, apps, computers, and tablets provide students with rich opportunities to support their learning and development in a range of ways. Through increased access to digital technologies, students benefit from learning that is interactive, collaborative, personalised, engaging, and transformative. Digital technologies empower our students to interact with and create high quality content, resources, and tools. This also significantly increases the scope for personalised learning, transforms assessment, reporting and feedback, and drives new forms of collaboration and communication."[58]

The restructuring of schools around digital infrastructures does not respond to pedagogical, student, or community demands, but to the corporate-inspired techno-capitalist imaginary that has positioned buzzwords like *information*, *digital*, and *innovation* at the center of basic education. Do we really need to sacrifice decades of educational efforts, teaching experience, and pedagogical experimentation on the technological and financialized altar of the gods of the digital economy? I am no expert, but from my precarious experience as a parent and teacher I believe there are many ways students could engage in learning that is "interactive, collaborative, personalized, engaging, and transformative" without having to be plugged into their laptops in every classroom as early as third grade (age eight).

The naturalization of the digitalization of classrooms represents, perhaps, one of the greatest political successes of large technology companies, although this achievement has not come without a cost. Under the slogan "Helping to expand learning for all," Google launched a global advertising and organizational strategy with the aim of winning the trust and loyalty of educators in its EdTech revolution. On its website, the corporation presents itself as a universal educational philanthropist, ready to provide the world with the tools and resources needed to unleash students' potential, ensure their online safety, and support Indigenous communities, among other goals.[59]

More importantly, the Google Teaching Center operates as a space to reeducate, evangelize, and help socialize a new generation of teacher-innovators through "free courses" designed to teach teachers how to navigate the Google sphere. Once completed, educators are certified by the corporation. Depending on the "level" they reach, educators can become "Trainers, Coaches or Innovators." In Google's courses, teachers can learn how to "explore the benefits of the digital classroom and foster 21st century work habits." It would be redundant to say that, for Google, "Getting Ready to Use Technology in the Classroom" (the title of the first unit of its introductory course) quickly equates to "Getting Familiar with Google Tools for the Digital Classroom."[60] This uncritical content, full of Silicon Valley hype, is presented as a cutting-edge pedagogical methodology with academic value.

In addition, the corporation also supports and funds Google Educator Groups. Billed as self-managed and self-organized communities of teaching professionals, they operate on every continent. These corporate spaces purportedly "bring local educators together, in person and online, to provide a forum for sharing, collaborating, and supporting each other in using technology in impactful ways with students."[61] While some of these groups operate with a low profile, others, such as those in Spain, act as lobbying groups coordinating campaigns, organizing courses for professionals, and joining forces with other private sector actors willing to accelerate the "digitalization of education."[62] Their operations are disguised as casual, informal, and disinterested acts: "It's not about looking for experts, but simply to share the practices we do in the classroom and at school in order to enrich each other. We would like to welcome teachers, management teams, people from the digital coordinations, companies. . . . Basically, people related to education and who use or want to use digital technology in the classroom to improve methodologies."⁶³ However, these events result in the cathartic celebration of Google's corporate power, with corporate education executives officiating inspirational speeches to the greater glory of digital capitalism.

The control exercised by big tech over the digitized classroom leads to a worrying realignment of public education according to private interests. An illustrative example can be found in the highly organized Google Education Group in Spain, which prominently features on its website a "Responsible Guide for the Google-based Digital Education Ecosystem."⁶⁴ However, rather than being a genuine guide, it serves as a tool for Google indoctrination. The guide is divided into sections aimed at administrators, families, and educators, offering information ranging from technical aspects to promoting the virtues and adventures of Google's philosophy. Notably, the section devoted to families features misleading templates of consent forms containing prefilled authorizations for both the Google ecosystem and other supposedly "necessary" third parties. These consent forms lack informational value, failing to provide clarity on how a minor's data will be stored or processed. Instead, they include a blank space to copy and paste a link to the corporation's terms and conditions page, further obscuring the true nature of the data use and privacy implications. This manipulation of consent forms and lack of transparency highlights the asymmetric power dynamic between big tech companies like Google and users, particularly students and their families. It reflects a disturbing trend where private corporations assert their control over educational ecosystems while evading important ethical considerations and regulatory safeguards.

As recent research shows, the digitized classroom favors digital enclosure of students. Teachers are provided with tools to track, monitor, and spy on students' online activities in real time, naturalizing a panopticon regime in classrooms.⁶⁵ Moreover, students are socialized in a context where granular surveillance is normalized. Meanwhile, corporations profit from and trade on the legal and illegal tracking of data in digitized classrooms.⁶⁶ The surveillance apparatus deployed on students is onerous and outrageous in its own right, but it also amplifies latent inequality. As research is beginning to show, those better off (with the means to purchase their own devices) are less surveilled and monitored. Chromebooks provided by schools often contain tracking software and hardware used to monitor students in and out of school hours. Of course, there are strong arguments regarding online harm that may justify some degree of content moderation.[67] However, the lack of clear regulation, transparency, and communication between students, families, staff, and corporations is leading to a dangerous segregation of student surveillance.[68]

The datafication of the classroom, which, as we recall, is driven by the neoliberal addiction to quantification, is impacting how the relationship between education and learning is interpreted. A significant example of this is the emphasis that different software programs place on trying to measure what is conceptualized as student engagement with the course. Often, the arbitrary and questionable metrics employed pay attention to factors such as the duration of online connections or the number of messages sent and received. Thus, quantity is prioritized over quality, neglecting important aspects such as the human experience within the classroom and the valuable exchange of knowledge between peers and teachers.[69] Under this paradigm, a student who evolves positively is one who surpasses different numerical thresholds that are preset and counted algorithmically. A process that constructs the educational process in the manner of a role-playing game that requires obtaining a certain score to pass to the next level. Evidently, algorithmic education does not manage to capture the nature, or the greatness if you prefer, of the vast and unbounded concept of education, which it reduces to a process of individualized quantification.

This underscores the paradoxes inherent in digital technologies. While they can challenge outdated practices that reduce learning to regurgitating memorized information on exams, they also give rise to new forms of dehumanized and mechanized learning. It is crucial to recognize the risks associated with datafied, competitive, hyperindividualized, and gamified classrooms, such as those offered by platforms like Class Dojo. These platforms allow teachers to use big data insights to make shallow psychosocial predictions about students. In addition, they provide behavioral mechanisms that allow teachers to shape student behaviors to conform to predefined notions of acceptable behavior within schools.[70] As can be seen the datafied classroom shares goals with the traditional disciplinary school: norm, control, modulation of behavior. But there is more. The technological leap allows the intensification, acceleration, and augmentation of old techniques to which it adds new strategies based on granularity. And perhaps more importantly, although questionable, the disciplinary school conformed to principles and objectives that, although vague and certainly erroneous, were aimed at the common interest. In contrast, the datafied classroom instrumentalizes students and teachers in pursuit of its ultimate goal: to generate economic benefits for digital capitalists.[71]

Families, Judges, and Criminal Corporations

In an unexpected turn of events that captured international attention, the quiet town of Helsingør, Denmark, became the epicenter of a global digital battlefield. In this scenario, a motley crew of actors including anxious and concerned families, neoliberal-leaning public authorities, and predatory technology corporations found themselves engaged in a full-scale asymmetric struggle. The tension culminated on August 19, 2022, when the Danish privacy watchdog, Datatilsynet, made a momentous decision to announce a ban on educational technologies provided by Google in Helsingør primary schools. This measure, described in a detailed document (2022), was based on the finding that the municipality had failed to fulfill its fundamental responsibilities as a data manager under the European Union's General Data Protection Regulation.[72] The authority stressed that the municipality had failed to adequately document and assess the inherent risks associated with the transfer of student data to a third country deemed insecure, in this case, the United States, through the use of Google technologies (Article 35 of the General Data Protection Regulation). The decision highlighted concerns about the security and privacy of students' personal information, an issue of growing relevance in a world increasingly interconnected and dependent on digital solutions.

As in many other regions, and like in the Canberra case, Helsingør's primary schools have gradually adopted digital tools in their administrative and educational operations.[73] These resources have become essential elements of daily life in schools, providing convenience in various tasks but also resulting in an increasing reliance on large technology corporations. Currently, many critical functions in schools require the use of specific platforms, forcing students to register and accept their terms and conditions, something that often takes place in a context of total informational asymmetry and corporate obfuscation. Initially, these corporations claimed that they would only use the data for educational purposes and that their "free services" did not harbor hidden agendas. However, recent events involving some of the most relevant actors (Google Classrom, Meta) have shown otherwise.[74] These educational tools have been found to have vulnerabilities that allowed third-party corporate partners unrestricted access to data. This raised concerns about privacy and consent, particularly given the company's history of handling personal data. But it was not privacy alone that led to the Danish authority's intervention.

Most public-private agreements in the field of digital transition in education have been marked by a lack of information, consent, and accountability.[75] This also applies to the way in which the accelerated introduction of technology into classrooms, if not the subsumption of the latter into the digital, has been implemented. Both students and parents, as well as the community at large, have been excluded from any consultation or explanation about the digitization of their educational spaces. Often, they have found that the mandatory use of technologies and digital devices in the classroom was a fait accompli, unappealable, a historical inertia. In this context, minors have been induced to enroll in digital services without parental knowledge or consent. It was precisely this scenario that triggered the case in a school in Helsingør. A parent concerned about his child signing up for YouTube for "academic purposes" without his consent filed a complaint with the Denmark's data protection regulator, Datatilsynet. This sparked a discussion locally and nationally about children's data rights and, more importantly, about the role and limits of digital platforms in the public education system. Of course, Helsingør's was not an isolated case.

In late 2019, growing discontent was echoed in Catalonia, Spain, when a coalition of parents, along with the prominent digital rights organization Xnet, raised their voices to the Catalan government. The concern lay in what they perceived as Google's deep and questionable meddling in the public school system. The tech giant, they argued, had not only integrated itself into teaching responsibilities and student assignments, it had also extended its reach into administrative functions. This situation, they pointed out, had been developed without due process of consultation with the educational community or observance of the relevant legal and administrative procedures. The concern was not minor: Google's ambiguous explanations about how it would handle and process children's personal data increased alarm among parents and representatives of civil society.[76] This unease not only sparked an intense political debate in the region but also mobilized the Barcelona City Council. The municipal entity, recognizing the seriousness of the situation, was quick to ally with Xnet. Together, they initiated the development of an alternative educational digital infrastructure, an experimental project that sought to offer more secure and transparent solutions.[77]

Meanwhile, in a broader context, Google was already facing international scrutiny. In particular, legal authorities in New Mexico were investigating the company for possible violations of children's privacy. New Mexico attorney general Hector Balderas took a strong position on the issue, highlighting a global landscape where data privacy concerns in education transcended borders and united different communities in a common cause:

> Google has also publicly promised never to mine students' data for its own commercial purposes. Unfortunately, Google has broken those promises and deliberately deceived parents and teachers about Google's commitment to children's privacy. In direct contradiction of its numerous assurances that it would protect children's privacy, Google has used Google Education to spy on New Mexico children and their families by collecting troves of their personal information including: Their physical locations; websites they visit; every search term they use in Google's search engine (and the results they click on); the videos they watch on YouTube; personal contacts lists; voice recordings; save passwords; and other behavioral information.[78]

Balderas not fooled by Google's narrative that it represented itself as a self-interested, philanthropic party providing free tools: "By tracking and cataloguing everything children do online and on their digital devices, Google has unprecedented visibility and access into the online lives of children across the country . . . though it is marketed to schools as purely educational tool, Google education provides far more benefit to Google than it does students or schools. Google recognizes that by giving children free access to its online tools and habituating them at a young age, Google obtains something much more valuable: generations of future customers." [79]

In the New Mexico case, as in many others, the outcome resulted in a settlement that appeared to have limited impact on the corporation, in contrast to the significant gains made. Google's director of government affairs and public policy expressed, with traditional corporate hypocrisy, "We are pleased to support programs and initiatives in New Mexico that promote children's education, privacy and online safety. We look forward to working with the Attorney General's Office to identify partners to help us execute this shared goal."[80]

An Extraction Machine

The privatizing spree of the 1980s that swept away public services and infrastructure in areas such as health, transportation, energy, water, and communications has become the new norm in education. Over the course of just fifteen years, large technology companies have succeeded in fulfilling the neoliberal dream of taking over public education and transforming it into a profit-oriented machine, creating a situation in which unscrupulous criminal corporations have become the vital infrastructure for millions of students and educators. This hacking would not have been possible in the absence of a sediment of looting, plundering, and undermining of common institutions blessed by the mainstream political spectrum: Liberals, Conservatives, Greens, and Labor.[81]

This process is vast, profound, and goes beyond a mere updating technology or a platforming educational structures. What we have before us is a paradigm shift in line with the designs of digital capitalism, a cyberpunk nightmare where teachers certified by digital corporations can monitor and control their students in real-time through the use of surveillance tools provided by financialized technological capitalism. This is a mechanism that, as I have shown, beyond the disciplinary, proposes an epistemological change in the meaning of knowing, learning, and teaching. In this sense, and supported by old exploitative logics, this transformation is making use of technological development to fragment the teaching and research experience, exacerbating the precariousness of academic work and reducing the experience of teachers to interchangeable and marketable packages, easily performed by temporary, expendable, and replaceable knowledge workers.

Beneath the facade of public-private partnerships lies the perverse logic of financialization, prioritizing profit over people. When financial capitalism infiltrates classrooms through start-ups, it undermines the very purpose of education—to foster progress and economic, social, and intellectual development—subverting it to serve the crude and brutal laws of the market, regardless of the rights at stake. This logic is injected into public services, seeking not only their commodification but also their complete reorganization under the dictates of capital. The same applies to the working conditions of teachers and researchers, a dedicated professional community that has historically resisted the neoliberal onslaught. Today, this collective finds itself in a dilemma: on the one hand, it is attracted by the power and relative comfort conferred by the new technologies; on the other hand, it is suffering their consequences.

The immediate repercussions of delegating critical educational infrastructure and responsibilities to criminal corporations are alarmingly evident. This process not only involves the massive transfer of resources and public wealth to wealthy and powerful corporate conglomerates, it also undermines the very possibility of building an alternative future. These processes have been carried out behind people's backs, with little or no information and with flawed consent. Many of the institutions have been and continue to be complicit in this process. Digital capitalism has established an unprecedented surveillance regime, where students and workers are constantly monitored in clear collision with the most elementary rights to privacy and the handling of personal data. Worse still, the datafication and digitalization of classrooms at all educational levels has accelerated the neoliberalization of the curriculum, making quantifying and unreflective logics the norm in countless cases. Moreover, the subordination of academic research to the opaque governance of knowledge, dictated by algorithms of digital capitalists, poses a serious challenge to the integrity and critical capacity of both researchers and universities. Education, historically a bastion of relatively free exploration and discovery, is at risk of succumbing to the envy once envisioned by neoliberals, placing it at the service of corporate interests and making its ultimate priority to generate profits and facilitate capital accumulation.

In the long term, the effects of entrusting the education of students, some as young as six years old, to profit-oriented corporations greedy for data and money, could be even more profound and disruptive. We are facing the risk that education, instead of being a path to personal and social development, becomes a tool for the creation of a highly controlled workforce shaped to meet market demands rather than fostering critical and innovative thinkers. The impact on the formation of values, the perception of the world, and the construction of identity of young people is a worrying unknown. The future of an education that nurtures curiosity, critical thinking, and creativity seems uncertain in this landscape dominated by technological capitalism.

3

Cyberwar against the People

On November 30, 2023, two Israeli media outlets, +972 and *Local Call*, revealed that Israel's military had made extensive use of an AI tool called the Gospel or "the word of God" to select targets for its bombardment of Gaza.[1] As the investigation highlights, the massive destruction of the Palestinian territories—where more than twenty-three thousand people had already been killed by January 2024—was not the consequence of indiscriminate shelling, but the result of a deliberate state terror strategy supported by AI tools, which helped identify targets for destruction. As one Israeli official noted in the report, "We are not Hamas. These are not random missiles. It's all intentional. We know exactly how much collateral damage there will be in each home."

This "mass murder factory" can create around one hundred targets per day divided into four strategic categories: tactical targets, such as troops or military buildings; subway targets, such as tunnels or hidden bases; "power" targets, such as energy, economic, and social infrastructures; and civilian targets, such as housing. According to Israel, the goal of this technological weapon is the destruction of Hamas's military and political potential, which it supports in two interconnected ways: (1) by assisting in the physical extermination of the more than thirty thousand people Israel has in its databases of people to kill; and (2) by trying to undermine the support of this organization among the Palestinian population. To this end, the Israeli army's AI vision has turned one of the most populated territories in the world into a military target susceptible to obliteration. Any space even vaguely related to Hamas—for example, the third floor of a ten-story residential building, where Hamas political officials meet—can become a military target subject to bombardment. In fact, most of the targets of Israeli attacks belong to the third and fourth strategic categories, which include hospitals, universities, schools, roads, and residential buildings, among many others. This AI weapon mirrors the logic of colonial extermination that drives this country, its very effects revealing the vastness of its targets: the genocide and dispossession of the Palestinian people. In other words, "the word of God" is an imperial weapon destined for war against the people.

There is no magic, no sorcery, no terminators. To generate these targets, the AI processes millions of data obtained from the surveillance networks to which Palestinians are subjected. The exact sources from which "the word of God" operates are not known with certainty, but Israel (as well as China and the United States) have used information from social networks, GPS, biometric data, censuses, records, bank accounts, and drone tracking in similar weapons for its cybernetic weaponry.[2] The substantial difference of the "the word of God" is its industrial capacity to generate targets in a way never seen before. How? Similar systems operate by establishing correlations, producing schemas, and generating social network maps, where, based on probability, a score is assigned to certain nodes.[3] That is, these systems establish who to kill, where to find them, and how much collateral damage is expected as a result.

"The word of God" does not identify subjects but evaluates targets based on the probability that a node is a terrorist, a threat, an enemy of the state—a statistical threshold that can make the difference between life and death not only for the node, but for everyone who crosses its path, including,more than ten thousand children, according to Save the Children. Where were the voices that preached the possibility of ethical Artificial Intelligences, the defenders of responsible technologies, those who said that with transparency and human control everything would be fine, everything would be safe? They were silent. They were silently being accomplices of a genocide supported by technologies such as AI. Unfortunately, this is not the only case in which states like Israel use state-of-the-art technological weaponry indiscriminately against subjects and groups considered dangerous.

Prior to the latest offensive against Palestine, an average of seventy thousand Palestinians passed daily through checkpoints in the West Bank and Gaza, where they were subjected to strict supervision by Israeli officials from both the public and private sectors. These Palestinians were treated as foreigners, enemies in their own land, being subjected to an exhaustive surveillance system that collected data from a myriad of sources, including fingerprints, iris scans, facial recognition, surveillance cameras, phone checks, social media spying, and so on. Many of them were arrested, accused of being adversaries of the State of Israel and imprisoned without trial under Israel's draconian anti-terrorism legislation. In other words, a legal kidnapping of the Indigenous population at the hands of occupation forces.[4]

In contrast to the rigid surveillance and separation imposed by physical and digital walls, these same borders are regularly crossed by Israeli occupation forces. In addition to conventional forms of supervision, control, and repression of the population, they carry out spying operations on phones and social networks to identify potential "terrorists," who are then eliminated, along with those close to them, through the use of drones. This system, highly systematized and rooted in Israel's colonial practices, led scholar Elia Zureik to characterize Palestine as a territory built by and for surveillance.[5] This is not a new phenomenon. Zureik's analysis traces this apparatus of surveillance and repression back to the genesis of the Israeli state, which already from its germ used tools of intelligence and violence at the starting point of massive dispossession of the Palestinian people, the Nakba of 1948. The Nakba, which involved the forced displacement of hundreds of thousands of Palestinians, and the subsequent occupation of their lands and homes by Israeli settlers, was meticulously planned with spying techniques, flights, photographs, and the use of advanced weaponry. Since then, the state of Israel has persevered in the sophistication of these methods (e.g., creating some of the first drone models as early as the 1960s) until solidifying an apartheid regime firmly supported by an assemblage of technological weaponry.

This situation is tragic in itself. But there is more. If the United States or Germany were to buy technology such as Gospel from Israel and use it to defend national security, they would not be in breach of any treaty, they would not be violating any agreement, despite these being technologies designed for extermination, programmed for destruction, and tested on tens of thousands of civilian bodies. Many of these technologies based on military artificial intelligence are in conformity with much of the global legislation, including its ethical and moral criteria and international political commitments regarding the laws of war. In other words, this weaponry would be a legitimate product. Unfortunately this is not a hypothesis; these weapons born of dispossession and extermination circulate globally.

Israel is one of the world leaders in the technological arms industry and military AI, as evidenced by its drones and software that are bought by half the world. The flourishing of this industry of death is one of the key explanations for the success of Israeli digital capitalism, both deeply benefited by the close involvement of the state with the country's main technology companies in the "digital occupation" of Palestine.[6] It is a key sector that not only generates annual revenues of billions of dollars but also allows Israel to exercise a diplomacy based on the export of death with countries such as Saudi Arabia or Morocco. Political and economic success is built on hundreds of thousands of corpses in this military experimentation field that Israel has turned Palestine into.[7] The technologies of the new Nakba flow following the rivers of global capitalism contributing to other processes of dispossession and colonialism.

Thousands of miles away, the United States is erecting a militarized physical and digital wall reinforcing the more than three thousand kilometers of colonial border imposed against Mexicans and dozens of Indigenous peoples. This wall is conceived from the ethics of rejection, fear, and racism and designed to prevent the arrival of undesirable and "invading" illegal immigrants. The sentinels of this wall are extraordinarily diverse. Some are as big as buildings, such as the more than three hundred highly technological towers (many of them fully autonomous) capable of detecting people day and night from miles away. There are small ones like household appliances, such as the automated license plate readers, which can register the license plates of any car, associate it with owners, establish patterns of use, and track it through its network. There are mobile sentinels, such as the hundreds of drones that fly day and night and the camera systems mounted on cars patrolling the border. There are invisible guards, such as the software that spies on and hacks migrants' phones and that makes it possible to scrutinize the huge databases and establish the correlations needed to identify and locate people from supposedly unidentifiable aggregate information.[8] This wall is built with national tech giants like Amazon and Palantir, military contractors like Anduril Industries and Lockheed Martin, and with experts in repression and settler colonialism like Israel's Elbit Systems, who entered the US market boasting of its surveillance expertise in the West Bank.[9]

The aforementioned cases are only a sample of a broader phenomenon, where the civil and the military, the public and the private, are diluted. They represesent regime of global dimensions with an unprecedented capillarity and a process that reconfigures neoliberal mechanisms to control, punish, and govern the poor—the misnamed wars on drugs, terrorism, and illegal immigration—in light of the logics of digital capitalism and technologies such as AI. In short, we are witnessing a technological breakthrough aimed at expanding antipopulation warfare operations on an unprecedented scale.

Cyberwarfare and the Responsible Military Use of Military AI

In the last ten years, the use of digital technologies in the military field has experienced an exponential increase, impacting multiple areas such as surveillance, logistics, and the development of new offensive aerial, land, maritime, and cyber strategies. Probably one of the most relevant and impactful examples are autonomous drones, profusely used by the United States in its "war on terror." Thanks to technologies such as AI, these weapons can autonomously perform spying and attack tasks, ranging from "surgical" to massive bombing. For example, in Iraq, drones have been widely used in operations against the so-called Islamic State (ISIS), playing a crucial role in assassinating key leaders of this group and reducing its operational capability.[10] Another example of the rapid transformation in the way warfare actions are understood and implemented has to do with the development of cyber weapons. Despite their intangible nature, these weapons can affect and destroy physical infrastructures. One of the best-known cases is the Stuxnet virus. This malware, developed in 2009 as a result of collaboration between the US and Israeli armies, was intended to infiltrate the industrial computers controlling the mechanical processes of Iran's experimental atomic power plants in order to affect and damage critical processes.[11] Their success brought about the obsolescence of the traditional distinction between cyber and physical weapons.

The emergence of new military capabilities in a global context where the digital has merged with the social through technologies such as the Internet of Things seem to confirm the suspicions of collectives such as Tiqqun, which spoke of a nomadic transformation of military strategy.[12] The world has become a global imperial scenario where any node, individual, or system can be immediately and remotely reached. This has been evident in conflicts such as those in Ukraine and Gaza, where the use of drone offensives, AI like Gospel, cyber attacks, and digital information censorship along with the mobilization of hundreds of thousands of boots has highlighted how the boundaries between physical and digital conflicts are completely blurred, marking a new era in the conception and execution of military operations.

The concept that has traditionally served as an umbrella for the multiplication of digital and information technologies in the field of warfare is that of cyberwarfare. It can certainly not be said to be an understudied concept. By consulting a main academic search engine in January 2024, I found more than thirty-thousand academic publications (thirteen thousand since 2021) containing this concept. And this does not include the countless reports, newspaper articles, and documentaries or vast array of books ranging from institutional propaganda to quasi-realistic science fiction. Despite this abundance of publications, the intellectual ecosystem around this notion is dominated by a fundamentally conservative view established in the early 1990s by analysts linked to the US Army. For example, a publication of the RAND corporation, perhaps the most influential think tank in the US Army, stated in 1993:

> Cyberwar refers to conducting, and preparing to conduct, military operations according to information-related principles. It means disrupting if not destroying the information and communications systems, broadly defined to include even military culture, on which an adversary relies in order to "know" itself: who it is, where it is, what it can do when, why it is fighting, which threats to counter first, etc. It means trying to know all about an adversary while keeping it from knowing much about oneself. It means turning the "balance of information and knowledge" in one's favor, especially if the balance of forces is not. It means using knowledge so that less capital and labor may have to be expended.[13]

Early attempts to describe virtual warfare scenarios reduced the concept of cyberattack to the use of digital weapons by state or nonstate actors within, on, or against political entities.[14] These early contributions also emphasized the distinction between cyber and informational offensives with respect to more traditional forms of kinetic or physical aggression. However, this did not last long.

The rapid evolution of cyberweapons such as the Stuxnet virus mentioned above ended this duality. More recent academic contributions have reflected the rapid transition to hybrid warfare scenarios, recognizing, for example, the duality of weaponry such as drones coexisting in informational and physical environments.[15] This trend accelerated during the recent armed conflicts in Syria (2011–present) and Ukraine (2014–present), where tanks and soldiers have been as relevant as drones, hackers, and disinformation campaigns. As the Australian Information Warfare Division's promotional video indicates, "Conflict is no longer limited to air, land and sea. The information environment is the new battleground. In this rapidly evolving world, we need to be prepared to fight with the computer as well as with a weapon, using armed information systems as much as armed vehicles."[16] Cyberspace has become not only the fifth domain of warfare after land, sea, air, and space but a threshold that intersects and captures all of them. At the micro level, the cyborg is no longer a poststructuralist figure of speech describing "a condensed image of imagination and material reality, the two conjoined centers that structure any possibility of historical transformation."[17] The era of the cyborg has reached its maturity, manifesting itself in some of today's critical military roles, such as piloting or analyzing intelligence data, functions under which mechanics, biology, and software merge. This amalgamation of disciplines transcends the realm of military intelligence, permeating naval, air, and infantry operations under the broad umbrella of cybernetics.

The rise of surveillance as a form of power based on information control is redefining the way governments understand national security. This is not an entirely new phenomenon; it dates back to the counterintelligence programs of the 1960s and 1970s against domestic "radicalism,"[18] a trend subsequently accelerated by the US Patriot Act of 2001. However, it was not until the administration of democratic president Barack Obama that the United States adopted a determined, comprehensive, and aggressive approach to cyberwarfare.[19] During Obama's tenure, significant efforts were undertaken to consolidate and strengthen the country's cyber capabilities. This involved the unification of several dispersed institutional initiatives under the umbrella of the National Security Agency (NSA). The blending of military and civilian affairs under the cyber surveillance umbrella has persisted during the Donald Trump and Joe Biden administrations during which information extraction and processing capabilities were further expanded through the creation of numerous data fusion centers aimed at protecting the country's national security and cyber interests. In other words, the Obama administration not only consolidated the US military-industrial-vigilante complex but fully integrated it with the country's cyber strategy by blurring offense from defense, military from policing, something that has been replicated in the rest of the Global North.[20]

For example, the leading European digital rights advocacy organization, the European Digital Rights network (EDRi), has repeatedly denounced the increasing militarization of European borders and the extensive use of cyber weaponry, ranging from drones to predictive tools used by Frontex, the European Border and Coast Guard Agency.[21] Other jurisdictions have also established collaborations between military, civilian, and private actors for surveillance and monitoring of online communications, as is the case with the UK's 77th Brigade and Israel's Unit 8200, whose operations range from developing powerful cyberweapons to creating viral content for social networks.[22]

Despite the growing global importance of cyberwarfare, its regulation remains in the dark within the framework of conventional laws of war. However, various initiatives, both at the national and regional levels, have emerged with the aim of defining and detailing key concepts such as cybersecurity, cyber aggression, and cyber diplomacy. For example, the United Nations has been actively engaged in discussions on norms of responsible state behavior in cyberspace seeking to establish an international consensus on how existing international laws, including the laws of armed conflict, apply in cyberspace.

In this context, NATO's Cooperative Cyber Defense Center of Excellence has played a significant role, notably by convening an independent group of experts between 2009 and 2012, whose work culminated in the creation of the *Tallinn Manual on the International Law Applicable to Cyberwarfare*.[23] It is an influential text, but one without binding legal authority. The work, currently in its second edition and known as *Tallinn Manual 2.0* (the next one is on the way), focuses on cyber operations and provides detailed guidance on the application of the principles of the law of war in cyberspace. It covers topics such as state sovereignty, state responsibility, use of force, and protection of civilians, and is considered an essential resource in the field of international cyber law.

The *Tallinn Manual*, while avoiding polemical definitions, introduces two fundamental pillars of the post-Westphalian order in the cybernetic era: the notions of state sovereignty and legitimacy and a Eurocentric perspective on international law. In their approach, they restrict the use of force to the doctrine of a just cause for war, a concept originating in the sixteenth century and deeply influenced by the School of Salamanca in the context of the conquest of America and Indigenous genocide. This doctrine has historically been used to justify imperialist and colonialist practices, providing an ethical and legal framework used to cover up economic and political interests under the guise of civilization or religion, something that has been brilliantly pointed out by critics such as Ntina Tzouvala.[24] Furthermore, the manual upholds the Weberian notion of the state as a central actor in the "legitimate use of violence," reinforcing the importance of the nation-state in the context of cyber belligerence.

In recent years, there has been a significant shift in the way cyberwarfare is conceptualized, particularly in those areas most closely linked to what is known as military artificial intelligence. In the past, discussions tended to focus on technical, strategic, and tactical aspects such as cybersecurity, cyber espionage, and critical infrastructure protection. However, as artificial intelligence has taken on a greater role, attention has shifted to ethical issues in what has come to be known as a whole new field of the moral philosophy of warfare.

The great debates of military ethical AI do not go into the imperial, colonial, or racist background of conflicts. Most of the literature respects the "legitimate" right of countries to defend (and attack) themselves in a tacit acceptance of the old state-centric and colonial-imperial "just war" discourse. They are not concerned, for example, with the spurious reasons that led the United States to invade Iraq or the current genocide in Gaza. For mainstream voices within this field, the problem lies in how to establish limits and safeguards that guarantee an "ethical and responsible" use of AI in military contexts, for example, by creating transparency mechanisms in the processes and mechanisms that allow for accountability in cases of errors and collateral damage. They are driven by the ethical, moral, and technical minutiae that derive from fatal human-machine interaction. In short, they are driven by questions such as: When, how, and why would the use of autonomous and semiautonomous weapons be justified? How, when, and by whom should they be regulated? A myriad of solutions have been proposed, from subjecting these weapons to humanitarian law "by code" to the implementation of more or less complex protocols, all with the philanthropic aim of creating "ethical" weapons that contribute to solving humanitarian crises.[25] Among the diversity of proposals there are certain common demands coinciding with the Eurocentric and liberal anxieties of individualistic capitalism: the need to include the "human" in the decision and the need for the system to avoid algorithmic biases. With that, it would seem, any ethical doubts about the massacre would be, at least from the point of view of military AI, resolved.

In this context, in November 2023, the United States made public the "Political Declaration on the Responsible Military Use of Artificial Intelligence," which has since been joined by some thirty countries. The document attempts to address the existing mainstream academic criticisms and suspicions about the use of military AI.[26] In this sense, the countries signing the document committed to maintain human ultimate control over automated decisions; to monitor for algorithmic bias; to ensure that the weapons are fit for the purposes for which they are designed (to crush the enemy); that they are programmed in accordance with the principles of international humanitarian law; and that their development is documented in order to be auditable and transparent. If these requirements were met, everything would seem to be solved—or do States not have the legitimate right to defend themselves against terrorist attacks?

Regardless of their practical applicability, documents such as the *Tallinn Manual on the International Law Applicable to Cyberwarfare* or the "Political Declaration on the Responsible Military Use of Artificial Intelligence" shed light on the prevailing legal and political discourses through which states (and their corporate partners) justify their actions and enable large-scale organized violence. For example, on the one hand, the "Political Declaration on the Responsible Military Use of Artificial Intelligence" endorses a traditional imperial view of international law where the technicalities of warfare are regulated without questioning the legitimacy of the imperial order that promotes them. On the other, it fills a shameful gap on the use of AI in defense and security matters, which, as many social and digital rights protection organizations have pointed out, have been conveniently excluded from brand new and useless legislation and legislative orders such as the European AI law.[27] It does so by rhetorically defending a progressive and responsible vision, in this case, through "the prohibition of autonomous weapons" and the "regulation in line with international humanitarian law of semi-autonomous systems," while opening the door to the development of the worst uses of these technologies. With these displays of legal fetishization, states establish and strengthen a framework of legal impunity for espionage, mass surveillance, hacking, disruptive activities, disinformation campaigns, and remote drone warfare. Yet these same states arrogate to themselves the sole authority to interpret the meaning of war and cyberwar, conveniently excluding from this definition devastating harm inflicted on individuals. Moreover, these documents reflect a power structure where dominant states and corporations define the rules of war and cyberwarfare, marginalizing and excluding the voices of affected communities.

It is clear that the one-dimensional conception, which simplifies (cyber)wars as mere confrontations between states or state structures, is, in the first place, insufficient to capture the social damage perpetrated by the use of new technological weapons. This restricted view also prevents us from situating contemporary transfers of knowledge and repressive technologies from the military to the civilian field where they belong, that is, in the long history of dispossession, violence, and imperialism. It is therefore imperative to resituate the terms of the discussion in such a way as to discern and understand how much of the latest developments in cyber weapons end up being implemented on the traditional targets of the state and capital: deviants, migrants, refugees, criminals, seditionists, and terrorists. And that is precisely what I propose to do in the following section: to lay the theoretical foundations necessary to offer a critical definition of cyberwarfare away from the traditional Weberian and state-centric parameters and provide a definition that allows its identification, even when official statements point to certain actions as police tasks, antiterrorist operations, or simply business as usual. To this end, I will draw on a new wave of critical scholars who, by paying attention to different structures of oppression—such as race, class, and gender— unravel the complex and colonial nature of cyberwarfare.

Cyberwarfare as a Historical Process of Dispossession

Drawing on the Marxist tradition, Nick Dyer-Witheford and Svitlana Matviyenko distinguish in their book, *Cyberwar and Revolution*, between three different understandings or dimensions of war: war as competition between capitalist states, war as conflict between capital and labor (or, in Marxist terms, class struggle), and revolutionary war, understood as a collective struggle for emancipation.[28] These three dimensions are neither exclusive nor pure and often intertwine and collapse into each other. For example, the US-Vietnam war (1954–75) was at the same time a clash between sovereign states, a colonial offensive, a class and civil war (both in Vietnam and the US), and an anti-imperialist revolutionary war for the liberation of the Democratic Republic of Vietnam. Each dimension not only describes a set of phenomena, but also implies a methodological approach that privileges some research priorities over others. For example, the first view focuses on states and their institutions. It is the one that has traditionally prevailed in academic circles and which I have briefly outlined above. My view is situated in the second and third dimensions. By emphasizing not the belligerent actors, but those who suffer, those who are dominated, those who are dispossessed, those who resist, this type of analysis allows for a quality exposure of the overlapping power structures imbricated in historical processes such as wars. It is not (exclusively) about criticism or denunciation, it is an epistemological exercise. Notions such as class struggle, imperialism, colonialism, and racism allow us to identify and point out "war" where mainstream science only sees disorder, justice, or anti-terrorism work.

In a sharp book, Roberto J. González has outlined a precise analysis of the devastating social consequences of militarization, datafication, disinformation, and predictive technologies.[29] His Foucauldian approach traces a genealogy of cyberwarfare by analyzing the securitarian-ideological interconnection between the Pentagon and Silicon Valley. Going beyond theoretical analysis, González lays bare the materiality of this relationship by exposing the immense transfer of public wealth to technological giants for the sole purpose of developing and strengthening a hypertechnologized militaristic leviathan. González focuses his analysis on the disinformation campaigns perpetrated by large corporations. He focuses on the Facebook–Cambridge Analytica scandal, which he rightly analyzes from the perspective of psychological warfare. In a particularly interesting approach, González highlights the close collaboration and camaraderie between corporations and armed forces in cyberwarfare operations: "In this virtual battleground, social media firms have laid the technical groundwork for deceptive, hostile messaging that has been largely uncontrolled and unregulated across much of the world. Silicon Valley is deeply implicated, to the extent that its firms have built this architecture, while its corporate executives have vehemently opposed government regulation."[30]

Throughout his book González stresses the relevance and impact of socio-technical predictive tools, which are used by intelligence and law enforcement agencies operating in both national and international contexts. In addition, he emphasizes the crucial role of the Defense Advanced Research Projects Agency (DARPA) in the funding and development of these data-driven technologies. DARPA has not only pioneered the creation of surveillance and data analysis tools, it has also facilitated their adaptation and application in various "battlefields," transcending the boundaries between national and international security. González's research reveals the similarities and material lines of continuity between the counterinsurgency tools implemented by US military forces in scenarios such as Afghanistan and Iraq, exemplified in systems such as Nexus 7 and W-ICEWS, and the technologies later employed in the field of homeland security in the United States, such as PredPol (which I will study later). This convergence signals a transfer and adaptation of repressive methods and strategies from the military theater of operations to the civilian management of public order. This dynamic highlights how technologies originally designed for military contexts subsequently find applications in the civilian security arena, evidencing a fusion and expansion of repressive tactics across a wide range of institutions, both military and civilian, and extending into state and corporate domains.

Brian Jefferson explores the use of military tech in civilian settings by focusing on the intersection of racial and digital capitalism evidenced by the rise of prison technologies.[31] His work traces the origin of today's "fully automated police apparatus" to the meeting of two phenomena in the 1970s: the conservative revolution in criminal policy that would result in a regime of mass incarceration and the computerization of police departments based on budget cuts and a neoliberal notion of efficiency. From the humble beginnings of data interpretation to today's aspirations for preventative crime, technologized socio-technical prison systems have become "powerful tools for invisibilizing and normalizing the methods by which cities administer racial criminalization."[32]

Abolitionist activist James Kilgore deepens Jefferson's critique, addressing the implementation of a "network of punitive technologies in the solution of social problems," which he calls "e-prison."[33] For Kilgore, this does not represent a different option to prison; on the contrary, it implies its expansion through the attachment of surveillance devices and restriction of freedom to the bodies of suspects, convicts, and parolees. Kilgore highlights how this digitized punitive regime has severe impacts on the lives of subjects and their communities, restricting fundamental aspects such as the ability to obtain employment, socialize and even access medical services. He also criticizes the ontological validity of reforms that, avoiding the debate on the abolition of prisons, propose their outsourcing and offshoring through repressive technologies.

According to Kilgore, the adoption of GPS tracking devices and other monitoring technologies in the criminal justice system represents a clear manifestation of a punitive prison culture that is focused more on control and surveillance than rehabilitation. These devices, which provide real-time location information, constitute an invasion of privacy and subject individuals under supervision to a state of constant surveillance, characterized by their inherently punitive nature. He also notes that the use of mobile apps for electronic searches and voice recognition by probation agencies is an extension of this control, limiting personal and collective liberty and perpetuating stigmatization. For Kilgore, digital curfews via GPS and breathalyzer and drug testing devices are restrictive control mechanisms that reflect the troubling expansion of prison logics beyond prisons. The self-proclaimed "former felon" criticizes how these practices infiltrate the daily lives of parolees, perpetuating a criminal justice system that privileges punishment and supervision over other mechanisms of justice. Kilgore strongly advocates resistance to this new digital panopticon, which only consolidates a punitive, carceral criminal justice system.

Brendan McQuade, for his part, delves into the ideological underpinnings of US data fusion centers, a heterogeneous array of data analytics infrastructures that supposedly serve to protect national security against the terrorist threat.[34] As McQuade explains, the US government has instrumentalized the so-called war on terror to conduct a global pacification campaign, thus blurring the boundaries between police and military power technologies. But what does appeasement mean? For McQuade and his colleagues in the anti-security school, Mark Neocleous and George Rigakos, it is nothing more than a continuation of the global colonial and social war of the ruling classes against poor, Black, racialized communities.[35] This interpretation helps explain the increasing militarization of police departments as well as the growing role of police in mass oversight functions, a phenomenon McQuade has termed the "mass oversight regime."[36]

In the extraordinarily well-documented *Borderland Circuitry*, Ana Muñiz explores the increasing technologization and multiplication of borders as well as the criminalization of Latino communities through their characterization as "gang members" and "illegal aliens."[37] Muñiz examines the immense surveillance information infrastructures that support repressive policies targeting Latino communities in the United States. She focuses specifically on two interconnected dimensions, immigration and gang surveillance, delving into the intricate details of these issues. As she demonstrates, the United States has created a costly surveillance circuit composed of federal, state, local, and private actors in order to manage a "dangerous" and "risky" population. Finally, Palestinian scholar Elia Zureik, who passed away in 2023, was among the most interesting and innovative voices in critical surveillance studies. In groundbreaking work such as "Settler Colonialism, Neoliberalism and Cyber Surveillance," Zureik describes Palestine as a territory fractured and reconditioned by Israel's surveillance practices.[38] Zureik's analysis highlights how seemingly different processes, such as endless conflict and occupation, Israel's start-up nationalism, neoliberalism, settler colonialism, and technological diplomacy, overlap to produce a regime of domination that institutes both the state of Israel and the pariah status of the Palestinian people.

Relying on these critical analyses, I conceptualize cyberwarfare against individuals as operations that, through the use of digital and hybrid weapons, generate or have the potential to generate a significant and socially harmful impact on noncombatant actors. This definition not only respects but also expands the scope of Article 8.2 on War Crimes of the Rome Statute of the International Criminal Court, integrating a broader and more contemporary perspective of warfare in the digital age. In this context, acts of cyberwarfare against individuals include a wide variety of offensive activities, ranging from surreptitious data collection to overt interruption of electronic, defensive, information, and communications systems. By this, I do not mean to revert to the legal fetishism mentioned previously but rather to provide an appropriate socio-legal framework from which to situate and identify these types of state and corporate crimes. Integrating this perspective into the analysis, I will devote the following section to succinctly examine three representative cases of cyberwar crimes against people perpetrated by state and corporate entities. In doing so, I hope to provide a detailed and illuminating insight into various modalities by which cyberwarfare operations—both state and corporate—inflict vast and routine social harm on hundreds of millions of people around the world.

The Digital Border Wall

Every year, thousands of people crammed into small boats, dinghies, and precarious vessels attempt to cross the invisible line that divides the European side of the Mediterranean Sea from its African coast, arguably one of the most unequal borders in the world. They take dangerous sea routes to avoid the digital border wall deployed by Frontex, the European Border and Coast Guard Agency.[39] Official statistics indicate that between 2014 and 2022, twenty-nine thousand people perished en route to Europe, but this is only a fraction of the people who went missing on their journey.[40] Frontex has become the EU's de facto first armed force and one of the most generously funded. For example, the Integrated Border Management Fund has received €7.37 billion for the period 2021–27,[41] much of which will be channeled to the state-corporate consortia developing Europe's digital border wall. It is certainly not the only such wall.

Thousands of miles away on the border that separates the United States and Mexico (and crosses the territories of numerous Indigenous nations such as Apache), hundreds of thousands of people face one of the most arduous and heavily guarded migration routes on the planet. Defying the dangers, these people take alternate paths in a desperate attempt to circumvent a digital border wall, a high-tech mesh under the watchful eye of US Customs and Border Protection (CBP). CBP's budget figures are a testament to this growing fortification: by fiscal year 2023, the allocated budget exceeded $25 billion, with considerable allocations made to border digitization.[42] This budget line item is a reflection of a progressive militarization of border surveillance, a process accelerated since the 1990s, characterized by hypertechnologization and partnership with defense contractors. Activists and scholars have come to call this phenomenon the "digital border wall," a vast military socio-technical structure that overcomes and extends the traditional boundaries of physical borders. The Mijente collective describes its heterogeneous composition:

> The digital border wall is made up of aerial drones, underground sensors, and surveillance towers amassed across hundreds of miles and capable of detecting humans, vehicles, and animals in all directions. It is the license plate scanners that catalogue every car in the border zone and the forensic kits that allow border patrol agents to retrieve personal data from these cars. It's the facial recognition, location tracking, and phone hacking tools available to a wide array of federal agencies operating in the borderlands. It is an attempt at total surveillance along the border and far into the interior, an effort by DHS [the US Department of Homeland Security] to monitor and control everything that happens between the United States and Mexico under the justification of border enforcement.[43]

Scholar and activist Petra Molnar distinguishes three dimensions of the digital border wall: before, during, and after crossing the EU's geopolitical borders.[44] The digital border begins before people on the move start their journeys, with data collection and automated risk assessment tools assessing the suitability of visa applicants from countries in the Global North. These processes increasingly involve automated and opaque systems operating as automated filtering, something that, as a recent Canadian parliamentary study points out, reinforces system bias, racism and discrimination in the processing of visa immigration applications by agencies such as Immigration, Refugees and Citizenship Canada.[45] Along with these technologies, other initiatives promote the use of drones in land, maritime, and aerial environments to track activities and communications related to illegal border crossings.

For example, the EU has invested considerably in the automation of its border control system. This was the case with the European ROBORDER initiative which proposed a swarm of unmanned robots with the aim of implementing "a fully functional autonomous border surveillance system."[46] This is certainly not an exceptional case. The EU has secured substantial funding for biometric data-hungry projects: €17.9 million for the Automated Border Control Gates for Europe and €9.3 million for the Robust Risk Detection System based on Pass and Baggage. These pilot programs are on the verge of becoming operational through the new entry and exit check-in system. Under the new scheme, the acceptance or rejection record of visa applicants will be stored together with their biometric data, thus becoming part of a vast digitized and interoperable European database. This new IT infrastructure will enhance an automated border control aimed at preventing irregular migration and helping protect the security of European citizens.[47]

The EU is also cofinancing the upgrading of security systems in Spain's infamous North African enclaves of Ceuta and Melilla. Melilla made international headlines in July 2022 for the killing of at least twenty-three people by police and the injury of seventy-three others (unofficial figures put the death toll at one hundred) while attempting to cross the southern European border. Described as carnage by the *Guardian*,[48] this is just one of many violent episodes that have occurred along Europe's perilous southern border. The already imposing cybernetic assembly of barbed wire, cameras, police, military, drones, and sensors of all kinds is being supplemented by facial recognition devices and other undisclosed technologies to be provided by leading Spanish surveillance company Indra.

Militarized border surveillance between the United States and Mexico is not limited to a simple physical barrier. It is composed of a complex network of advanced surveillance and enforcement systems. Among them are the automated license plate readers, installed at thirty-nine Border Patrol checkpoints, where they collect, date, time, GPS coordinates, and the license plates of vehicles crossing the border and then upload the information to an accessible database. Added to this is Immigration and Customs Enforcement's collaboration with Vigilant Solutions, a data trafficker, which has amassed more than five billion license plate records and 1.5 billion data points from more than eighty law enforcement agencies across the country. But the digital wall doesn't stop there. It includes Integrated Fixed Towers (IFTs), some as tall as 140 feet, equipped with cameras and radars capable of identifying people six miles away.[49] These towers, linked to the Torch-X command and control system developed by Elbit Systems initially for Israel's separation barrier in the West Bank, are only part of a larger surveillance system. This system also incorporates the Remote Video Surveillance System, which encompasses 368 towers distributed along the southwestern and northern US borders, and the Mobile Video Surveillance System, operated by Tactical Micro.[50] The technology reaches its pinnacle with Anduril Industries' Autonomous Surveillance Towers, which use artificial intelligence to identify and classify items of interest, differentiating between people and animals without direct human intervention. These towers, capable of autonomous and sustainable operation thanks to their solar power, represent a new era in border surveillance, where AI technology becomes an omnipresent and omnipotent actor.[51] In addition, CBP uses a growing fleet of drones, with more than 135 in operation and plans to acquire up to 460, operated by nearly six hundred trained agents and a goal to double this number.[52] Manufacturers such as AeroVironment, FLIR Systems, and Lockheed Martin, with multimillion dollar contracts, are at the forefront of this expansion. The digital wall also includes extensive biometric data collection. CBP agents, when apprehending an individual, collect not only DNA samples for the Federal Bureau of Investigation, but also fingerprints, facial photographs, and iris images, transmitted in real time to the Department of Homeland Security biometric database. This information is integrated into the Homeland Advanced Recognition Technology System (HART system), hosted on Amazon Web Services, which aggregates, links, and compares facial recognition images, DNA profiles, iris scans, fingerprints, and voice recordings from hundreds of millions of people.[53]

The digital barrier transcends the physical boundaries of the border, tracking those who cross them, monitoring their movements to workplaces, schools, and universities through a complex web of institutions, immigration and law enforcement agencies, and private corporate entities. A report entitled *American Dragnet Data-Driven Deportation in the 21st Century* by the Georgetown Center for Privacy and Technology reveals how Immigration and Customs Enforcement uses surveillance data infrastructures at the local, state, and federal levels, originally established as antiterrorism measures.[54] This report evidences how Immigration and Customs Enforcement, with the help of data brokers such as LexisNexis and surveillance contractors such as Palantir, cross-references data from multiple sources including criminal, educational, and health records, as well as driver's license and vehicle registration information, utility bills, and more, in order to identify potential subjects for removal from the country. These data-intensive operations have involved spending at least $2.8 billion between 2008 and 2021.

Beyond this digital frontier, controlled by tech giants such as Elbit Systems and Anduril Industries,[55] a vast coalition, including local and state police, tech giants, and powerful government agencies such as the NSA, conspires to launch a vast surveillance network. The goal of this industrial complex goes beyond collecting, processing, and analyzing large-scale data; it seeks to predict individual and collective behaviors, anticipating risks, crimes, and potential "national security threats." For example, US police departments, with the support of large technology companies, have established huge databases such as CalGang, an electronic arsenal designed to provide law enforcement agencies with accurate, timely, and electronically generated statewide information on gang-related activities.[56] This phenomenon can be interpreted as the use of automated blacklists to build civil, criminal, and deportation cases. As Ana Muñiz points out, these practices are a clear example of how new informational techniques merge with previous neoliberal punitive tendencies to criminalize immigration (a phenomenon often called *crimmigration*).[57]

This extensive surveillance network, supported by multibillion-dollar budgets and collaboration with defense contractors, reflects a growing militarization and a profound transformation in the nature of border surveillance. These technologies, far from being limited to immigration control measures, expand the power of the state to monitor, track, and control not only migrants but anyone within its jurisdiction, often without their knowledge or consent. This convergence of state initiatives marks an exponential growth in the state's ability to exercise pervasive control, raising critical questions about privacy, human rights, and the limits of state and corporate power in the digital age.

Surveillance, Assassinations, and Political Repression

In 2020, the *Guardian* uncovered a shocking fact: prominent Catalan politicians were victims of a cyberattack using the advanced Pegasus spyware.[58] Developed by Israeli defense firm NSO Group Technologies, this powerful software allowed attackers full access to their targets' cell phones, enabling call listening, location tracking, message interception, and even microphone and photo gallery use.[59] In essence, devices intended to facilitate communication and convenience became covert surveillance weapons. By 2022, the full extent of the scandal was fully revealed. Research by the Citizen Lab at the University of Toronto showed that at least sixty-five Catalan politicians, including the last four presidents of Catalonia, their teams, and, in some cases, their families, were spied on by the Spanish secret services.[60] Weeks after this revelation (April 18, 2022), the director of the Spanish spy agency resigned, although the government strongly defended the legality of political cyberespionage. This was not an isolated case. High-profile reports and hearings, from the European Union to the UN and the Inter-American Court of Human Rights, highlight that the Pegasus cyberweapon was used against political dissidents, human rights defenders, and journalists in countries as diverse as France, Hungary, Poland, India, Germany, Mexico, Yemen, and Saudi Arabia, among others.[61] For example, the UN investigation into the murder of journalist Jamal Khashoggi identified Pegasus as one of the tools used by the Saudi secret services for his extrajudicial killing.[62] Is this a recent phenomenon? Of course not. Despite the technological promises of transparency and democratization of media, the close relationship between datafication, espionage, and warfare has been a constant in digital capitalism.

On June 1, 2013, former NSA contractor Edward Snowden met with filmmaker and documentary filmmaker Laura Poitras, as well as *Guardian* journalist Glen Greenwald, at the Mira Hotel in Hong Kong. There, they nervously discussed how to disclose confidential information gathered by Snowden about one of the largest surveillance programs humanity has ever known. Four days later, the *Guardian* published the first of many articles describing the mass surveillance apparatus with which the giant NSA spied on the "networked" society.[63] This first report focused on leaked official US court documents that granted government agencies access to the phone records of Verizon, one of the largest US telecommunications providers, regardless of whether users were under investigation or not.[64] This was just the beginning. Documentation released later detailed the NSA's Planning tool for Resource Integration, Synchronization and Management (PRISM) program, a surveillance tool used by the agency to massively and indiscriminately collect information from electronic services provided by Apple, Google, Facebook, and Microsoft (among others).[65] The revelations exposed by Snowden depicted a dystopian narrative in which a vast surveillance agency, in collaboration with its counterparts in the other four of the "Five Eyes Alliance" (Canada, United Kingdom, Australia, and New Zealand), had unlimited access to the content and metadata of billions of individuals worldwide. This included an unprecedented volume of information, ranging from general aggregate data to highly specific details such as emails, photos, phone calls, and text messages.[66]

Previous instances of mass surveillance had been carried out by various political systems, including liberal, illiberal, and authoritarian regimes, spanning decades, if not centuries. However, the magnitude and implications of the scenario that unfolded marked a turning point in surveillance practices. In a matter of months, numerous terms such as "big data surveillance" and "datavigilance" emerged to capture the new reality characterized by the reversal of the burden of suspicion.[67] Suddenly, people around the world realized that their right to privacy had been revoked as they became subjects of a global, digitized inquisition orchestrated by agencies empowered to "kill based on metadata." While the PRISM program focused primarily on widespread communications interception, another program known as SKYNET focused on identifying and locating potential military targets, especially in the Global South.[68] Making use of a combination of techniques such as machine learning and quantitative social analysis on the Global System for Mobile Communications (GSM) metadata, location information, access, device types, and usage patterns, it identified digital suspects—information that would later be used to determine targets for the strike drone program that the United States used extensively in Afghanistan and Pakistan.[69]

The machinic production of the terrorist subject allowed the repressive structure to circumvent the liberal constraints of the right to due process, as the technologized system was not concerned with assessing the actions and words of a natural person to determine guilt. Instead, it assessed the risk of individuals identified as terrorists datafied to pose a threat to US interests. Accordingly, death sentences were regularly issued based on the statistical probability of possessing devices associated with "terrorists." This probability was inferred from factors such as the subject's proximity to other potential terrorists, residence in dangerous and hostile areas, and lack of a sufficient traceable record. Thus, being labeled a terrorist was the logical outcome of complex but verifiable quantification processes. This dehumanized regime of technologically assisted killing is the shared heritage of both the left and the right, regardless of the former's supposed ethical qualms.

In 2013, the *Guardian* revealed a confidential Obama-era document outlining the US Offensive Cyber Operations strategy, emphasizing the importance of identifying resources and targets for future operations while respecting international law, including sovereignty considerations and laws of armed conflict.[70] Despite this (or precisely because of it) during the Obama administration, drone operations against suspected terrorists were intensified, resulting in 324 civilian casualties out of the 3,797 people killed in 542 authorized drone operations. And these are just the acknowledged data.

Under the Trump administration, a significant change in drone policy was observed. Trump abolished the Obama-era approval system, which required hierarchical, high-level oversight, and replaced it with a more decentralized approach. This gave military and CIA officials the discretion to launch drone strikes against targets without White House approval, thereby reducing accountability for drone strikes. Trump also designated broader areas in Libya, Somalia, Syria, Yemen, and Iran as exempt from civilian casualty disclosure and repealed annual reporting requirements for civilian and enemy casualties in drone strikes outside war zones.[71] Predictably under the Trump administration, civilian casualties in drone strikes increased significantly. These attacks are part of a trend of increasing brutality and frequency of US military aggression first under Trump, then under the Joe Biden regime, demonstrating the lines of continuity in US digitized imperialism.

Numerous critiques have revealed the profound flaws and catastrophic effects of targeted and indiscriminate killings perpetrated by drones. Various voices point out how the digitization of imperialism—riddled with errors and developed from limited and biased datasets—operates under a veil of opacity and lack of accountability.[72] The problematic and lax methodology for labeling a person as a terrorist based merely on metadata has been emphasized.[73] On the other hand, Akbar Ahmed has highlighted the lack of contextual sensitivity in the application of a socio-technical system that overlooks the complex and shifting nature of tribal political and military alliances in regions such as Pakistan and Afghanistan.[74]

Drone attacks also have an alarming collateral casualty rate, in some cases as high as 90 percent. They not only take lives and disfigure bodies but also sow a legacy of collective trauma and fear in already devastated communities. There is a need for a critical examination of the ethics and effectiveness of these practices, which have inflicted enormous and unjustifiable suffering in the name of security and counterterrorism.

Military, Academics, and Police

In June 2022, a DARPA-funded University of Chicago project claimed to have created an algorithmic model capable of predicting the probability of crime with 90 percent accuracy. However, there was no magic behind it. The system processed geolocated and historical crime data from several US cities (Atlanta, Austin, Chicago, Detroit, Los Angeles, Philadelphia, and San Francisco) along with socioeconomic data sources (people living below the poverty line, unemployment rates, age, urban hardship index, housing types) to obtain spatiotemporal crime points.[75] Interestingly, the authors focused on a very specific type of crime: "We consider two broad categories of reported criminal offenses: violent crimes consisting of homicides, assaults, and batteries, and property crimes consisting of home invasions, burglaries, and vehicle thefts."[76] These are not the crimes that affect the majority of people nor those that cause the most social harm. Rather, these examples highlight white, bourgeois anxiety toward petty crimes against property and public health. A narrative traditionally instrumentalized to fuel ultraconservative criminal justice policies especially directed against the poor.[77] In other words, the research project used defense funds to create a technology explicitly targeting the most vulnerable people in the city of Chicago, which tripled the already high poverty rates of the state of Illinois.[78] This is an example of a tool in the service of predictive policing, a concept that encompasses a wide range of analytic technologies with two key characteristics in common. First, these are risk-based approaches designed to predict the likelihood of a criminal event occurring. This prediction is translated into a "score" for potentially dangerous subjects or crime-prone areas. When this score reaches a certain threshold, specific police protocols are activated, such as the deployment of patrols or notification of potential offenders. Second, to generate these risk predictions, these technologies rely on a large amount of data, which can be individual or collective, direct or indirect, historical or real-time, but which in any case elevates the quantification of bodies and statistical science to the category of predictive oracle.

The 2022 project draws from a long history; Chicago has long been known as a laboratory for experimenting with predictive policing initiatives. As early as 2009, the Chicago Police Department was already implementing experimental programs based on risk assessment with the goal of "reducing violence" through an intelligence-driven approach to fighting crime. These programs were presented as an alternative to conventional policing methods, emphasizing the use of community-based services to help people at risk of becoming involved in violent incidents. Between 2012 and 2016, the Chicago police department used a system to algorithmically generate and produce scores that reflected the likelihood of a person being involved in shootings.[79] The Chicago data portal proudly stated that their system did not use race as a data point; rather (among other things) it used "the number of times a person has been a victim of a shooting incident, age during the last arrest, number of times a person has been a victim of aggravated assault or assault, number of prior arrests for violent crimes, gang affiliation, number of prior narcotics arrests."[80] Based on these data, the system generated a secret "strategic subject list," which was distributed among the police ranks. Predictably, the "color-blind" algorithm ended up including about 56 percent of Chicago's young black male population (age twenty to twenty-nine) on these lists. Although race was not explicitly coded, all indicators associated with the racialized distribution of poverty fed into the algorithm. Further, the historically biased police actions contained in their files also served to feed the algorithm, a veritable breeding ground for racist technologies ready to target and signal black neighborhoods as a threat.

Notwithstanding initial claims that the project was a proactive social services approach, the system was used to increase police presence in areas that were already hypermonitored and to arrest high-scoring individuals.[81] However, despite the considerable resources invested in this techno-solutionary approach, Chicago experienced a 58 percent increase in homicides and 43 percent more shootings in 2016 (a year that recorded 764 homicides, making it the most violent to date).[82] A comprehensive review of the system, conducted as part of research funded by the RAND corporation, showed that use of the automated risk assessment tool did not result in a significant reduction in violence.[83] The project was eventually dismantled in 2019; however, as the DARPA-funded University of Chicago project described above shows, its spirit lives on.

Academic complicity in the transfer of racist military technologies to the civilian realm is a deeply disturbing and unfortunately common phenomenon. A paradigmatic example of this phenomenon is the case of PredPol, a crime prediction tool that was widely used by the Los Angeles Police Department (LAPD). This system was initially conceived by University of California, Los Angeles, anthropology professor Jeff Brantingham. Brantingham initially specialized in the archaeo-anthropological study of prehistoric populations. However, his behavioral methodology halfway between mathematical statistics and the social sciences brought him closer to the theories of criminal ecology, a subdiscipline of criminology that seeks to explain (and predict) criminal behavior based on the influence of the environment on subjects and societies. However, the course of his research took a momentous turn when he received a generous grant from the Pentagon, allowing him to significantly advance his work and, more worryingly, apply his findings in the civilian setting. Along with two other principal investigators, Brantingham led three postdoctoral researchers and six PhD students in the Dynamic Models of Insurgent Activity project.[84] The project, originally intended to determine geographic areas in Iraq with a higher probability of battlefield casualties, became a proprietary product under the for-profit company PredPol LLC, who successfully marketed it to a multitude of US police departments.

What Brantingham offered was a geographical crime prediction tool, commonly known as crime "hot spots." To achieve this, he used the processing of historical crime data streams categorized into several variables. PredPol generates maps that delineate five-hundred-by-five-hundred-foot areas, which are algorithmically identified as "hot spots" where crime is estimated to be most likely to occur in the future.[85] Like the Chicago tech system, the PredPol technology does not seek to predict white-collar crimes, such as ecological or economic crimes—that is, the type of crimes that the wealthy classes use to obtain and safeguard their privileged status. Following the neoliberal criminological hypothesis of "broken windows," PredPol promises to reinstate "order and law" in depressed and racialized neighborhoods by cracking down on crimes such as robberies and assaults (of the poor).[86]

The PredPol example, studied in several papers by, among others, the Stop LAPD Spying Coalition,[87] brutally exposes two defining characteristics of cyberwarfare against individuals. First, it shows the fluidity of relationships between public and private actors in the cyber defense and cyber policy industry as well as the critical role played by academia in the process. Second, and perhaps more disturbingly, it highlights how technologies designed to support highly questionable military operations, often directed against people of color abroad, are turned against people of color in the United States. In short, PredPol reveals how the criticized repressive capital and knowledge-power flows overlap with increasingly militaristic and punitive policies, thus forming a dangerous global circuit of repression and domination. As has been seen in a multitude of jurisdictions, this type of technology only reaffirms the racial and class prejudices of the law enforcement agencies that use it, since what it does is to establish predictions and correlations between existing databases, for example, by indicating which subjects may be more likely to commit terrorist acts, or which areas may be more likely to be a source of threats to national security. Needless to say, what is considered a threat, risk, or terrorism responds to the criteria set by law enforcement agencies. In other words, they serve as mechanisms for the scientific validation of repressive actions, also allowing the private sector to enter critical sectors of collective security and welfare.

Cyberwar or Just War?

The big arms lobbies, the great technological powers, and mainstream academia tell us that military artificial intelligence is inevitable and even necessary, but that it is necessary to regulate it. That is, they tell us to assume it as a fait accompli, but that we can make this technology moral and responsible, that we can design democratic weapons in accordance with international humanitarian law—the same law that smokes among the remains of the corpses of those murdered based on metadata. Governments around the world boast that AI is a tool for progress while entrusting the development of these military infrastructures to ultra-right-wing corporations notorious for their crimes and human rights violations. They speak of responsible and ethical artificial intelligence while drones of the Israeli Elbit Systems, tested in Palestine, patrol the Mediterranean under the orders of Frontex. What is this "responsibility" and "ethics" they are talking about? Do they mean some sort of ethical and responsible genocidal AI? Do they mean some kind of twisted ethics of walls and expulsions? The ethics of indiscriminate surveillance and spying? We must not fall into the trap of regulation, of normalizing destructive technologies. There are things we simply have to say no to. The only responsible, ethical, and safe military artificial intelligence is the one that does not exist.

This chapter has studied how the old dystopian dreams of cyberwarfare have become common practice in today's politics. However, contrary to what more traditional contributions show as technologies used by states to harm the interests of other states, the current use of cyber weapons targets the most vulnerable, political dissidents, migrants, refugees, and racialized and vulnerable communities. This revelation may come as a surprise to some, but it aligns with other manifestations of organized state violence. In these cases, the use of force by states is disproportionately directed towards civilian populations rather than engaging in conventional warfare against armed adversaries in conventional warfare scenarios.

Traditionally, the social sciences have largely ignored the idea of war beyond the narrow confines of Eurocentric, state-centered theories, neglecting to understand the colonial and racialized management of conquered populations for what it really was: war by other means. War is not limited to the massive mobilization of soldiers or the terrible effects of armies in conflict. War is also the constant surveillance of a dominated population, repression, cultural assimilation, organized rape, targeted assassinations, displacement, unjust distribution of wealth, the theft of entire generations of children. These are not the consequences of war, but the war against people itself. At this point (at least) one question arises.

What Is to Be Done?

A powerful example of worker solidarity and power was demonstrated in 2018 when Google employees protested their company's collaboration with the Pentagon on artificial intelligence drone imagery data analysis.[88] This display of collective action was followed by another notable instance in which Google workers protested their company's pursuit of the massive Joint Enterprise Defense Infrastructure (JEDI) cloud computing contract with the Pentagon. The workers' activism led Google to eventually withdraw from the controversial competition, leaving Microsoft and Amazon competing for the contract.[89] However, the fight continues. In May 2021, amid the worsening situation in the West Bank occupation of Palestine, Google and Amazon announced their participation in Nimbus, a $1.2 billion Israeli cloud project that plays a crucial role in expanding the digital occupation of Palestine. In response, Google workers, along with their colleagues at Amazon, persisted in protesting their respective companies' involvement in military affairs. They organized public events alongside Palestinian activists and launched an extensive information campaign. Despite their valiant efforts, these workers faced repression. For instance, 28 Google workers were fired on April 2024 after participating in protests again the project.[90] As one group aptly put it, "It has become impossible to express any opinion of disagreement with the war waged against the Palestinians without being called to a human resources meeting with the threat of reprisals."[91] However, the organized working class in technology has not been intimidated by fearmongering corporate repression. Instead, they have remained steadfast in their commitment to defy what is undoubtedly seen as collaboration with a brutal regime of organized terror. Their unwavering perseverance demonstrates the power of collective action and serves as an inspiration to others who seek to confront unjust practices and stand in solidarity with oppressed communities.

4

The Materiality of Digital Exploitation

We're fighting so our kids grow up in a society where they're not subjected to the exploitation of a big corporation like Amazon.
—Michelle Valentine Nieves, Amazon worker, union leader.[1]

It was a mild April afternoon in 2023. I was walking with my colleagues from the University of Melbourne to a seminar at the Royal Melbourne Institute of Technology (RMIT, a nearby university) when my gaze fell on a delivery man. He was standing on the side of the road, anxiously looking at his mobile phone, visibly agitated. My colleagues walked on; nothing seemed out of the ordinary. There are thousands of cyclists delivering meals to the *progressive* households of inner Melbourne. The overwhelming majority are young migrants, newcomers who are just starting to get to know the city. They come with temporary visas. They earn little and work hard. They suffer numerous accidents and usually bear the expenses of the necessary means of production to carry out the work from which others will benefit. I know this because I have had the opportunity to share political events and informal chats with rider's union organizers. I stopped and asked the rider if they were okay, or if they needed anything. I immediately recognized the beautiful Latin accent. After introducing ourselves, we continued the conversation in Spanish. His phone had stopped working—maybe the SIM card? He had a bike, the gear, the delivery backpack, and the food from the restaurant. He had the client's address and energy to pursue the task. However, he needed to be logged into the Uber Eats app to report his real time location and the status of the delivery. The app—legally an aiding software—was effectively his supervisor, his manager, and the capitalist tool for algorithmic exploitation. Daniel (this was his name) was going through a tough time economically. Moreover, his visa was expiring in three days, and immigration authorities were unresponsive.

He explained all this to me nervously, saying that he felt paralyzed and was unsure of what to do. "Do you know of any job?" asked. "No," I said, quickly realizing the privileges that my temporary position at the university granted me. I gave him the contacts of a couple of comrades, and we said goodbye. I never made it to the critical event on digital cultures that was taking place at RMIT.

During the long lockdowns of the Australian metropolis, delivery riders were praised for their work, exposing themselves when we still knew little or nothing about COVID-19. Suddenly, some of the stories behind the anonymous faces of those who delivered fast food to every corner of the cities in the changing and unpredictable Australian weather came to light. Thanks to investigative journalism and union complaints, the public were able to catch a glimpse of the riders' precarious lives.[2] The general audience got to know about the algorithmically imposed working conditions, as well as the risks, the accidents, and the deaths. The public learned about Chow Khai Shien (from Malaysia, twenty-six years old), who was hit by a car one night in 2020 while working for Doordash. They learned of the deaths a few weeks earlier of Uber Eats worker Dede Fredy (from Indonesia, thirty-six years old) and Hungry Panda worker Xiaojun Chen (from China, forty-three years old). The public learned that the families of the victims would not receive compensation for accidents and deaths, as stipulated in current legislation.[3] For the capitalists, these riders were part of a disposable, replaceable, and sacrificeable surplus population. According to the "gig economy" companies, the worker is not actually an employee, but rather an entrepreneur providing services to clients. These corporations claim to be merely the facilitators of a free exchange between parties and say that their role is limited to providing the necessary software for these services to be carried out (while cashing more than 25 percent of the transaction). This, as any Uber Eats, Deliveroo, or Doordash worker knows, is a lie.

Digital corporations are the ones determining every aspect of the working conditions, able to suspend workers' access to their platform arbitrarily—in other words, to fire them with no substantial explanation. Corporations evaluate workers while also controlling and moderating their relationship with clients; they define prices, set standards and procedures, decide the routes to follow, and, of course, control the company's narrative. For these companies, all these elements do not point to an employment relationship but to a contractual relationship between two private companies, regardless of the evident asymmetry that occurs, for example, between a delivery driver and Deliveroo. This lack of recognition of the worker's status is a blatant violation of the most fundamental labor rights, as workers have denounced and multiple court rulings have stressed.[4] As several international institutions point out with concern, surveillance at work has multiplied and become more granular.[5] Workers' steps, keystrokes, presence, and attention are measured by all sorts of smart sensors. Driving cadence, online messages, work rhythm, sent and received messages, and browsing history are controlled and registered. AI technology is used to evaluate images, videos, sounds, and temperature, all captured by countless data points. Under this algorithmically augmented Taylorism, digital capitalism is creating a whole new science of statistical control over workers, by measuring, observing, and managing everything in order to increase corporate power over work while multiplying their profits.[6] This is just a small sample of the criminal strategies that digital corporations use to exploit workers. Despite these aggressions, the working class resists, protests, and organizes. A testament to this is the thousands of Amazon workers who are challenging the company's antiunion tactics by consistently organizing and striking against the low wages, exploitative working conditions, and oppressive regime of mass surveillance.[7]

In the following pages, I will examine the ways in which digital capitalists are implementing a deeply racialized system of algorithmic exploitation. I will begin by exploring the system's historical and technological context, emphasizing the ways capitalism has historically relied on the exploitation of subaltern subjectivities. Next, I will explore the mechanisms by which digital capitalists exercise control over the means of production and the social fabric; I will look especially at the intersection of code, law, and digital capitalism. Finally, I will highlight some of the most visible social harms inflicted on the working class by this exploitative model.

The Missing Workers

When the labor performed by the *mythological* figure of industrialism—the white, male, heterosexual worker—was relocated from global centers to what were then peripheries, intellectuals suddenly found themselves without an explanatory framework for their reality. A horde of social scientists competed to be the first to certify the death of the working class (which they did in absentia, as the corpse remained elusive). Liberals, conservatives, socialists, third-way postmodernists, Marxists—everyone was present at the funeral, vying to officiate the ceremony. They sought not only to bid farewell to the past but to herald the new and brilliant post-work-centric future. The knowledge and immaterial worker, the cognitariat, and a bestiary of subjectivities emerged from the magical factory of postmodern sociology.[8] However, despite their tremendous efforts, digressions, and more or less grounded predictions, what lay ahead was undefined. Of course, not everyone fell for this farce. For instance, the brilliant Marxist intellectual George Caffentzis pointed out the blatant contradictions of the "end of work/worker" theories.[9]

The closest that postmodern sociology came to defining the new subjectivities emanating from this transition was when the (then archaic) concept of the precariat, rearticulated as precarity by Isabell Lorey,[10] acquired the indeterminate meaning it has today. Precarity was defined as a condition of vulnerability characterized by the end of the white and heteropatriarchal myth of Fordism: family, work, car, housing, and a predictable life trajectory. Precarity came to define everything that was outside the social protection regime. The concept of the precariat was born to define the growing otherness in capitalist societies: migrants, temporary workers, and renters. Precarity spread until it became the palpable reality of many, until it became the new normal.[11] Today, precarity is the state of existence for Millennials and the starting condition for Generation Z. The concept of the precariat may be useful for understanding elements of the social suffering emanating from neoliberal politics, but it does not serve to analyze the legal, economic, and technological mechanisms on which it is built. Precarious is the cry of the current working class, and perhaps it is their condition, but it does not serve to explain the logics operating behind the material conditions of their exploitation. Nor is it useful to understand the logics and tools of capital in the twenty-first century. Certainly, many workers disappeared from the privileged view of academics in the Global North, but labor multiplied.[12]

The idea that work is not confined to the narrow limits of the white male archetypical worker has always been present in feminist Marxism and critical race studies—two schools of thought key for understanding the current regime of exploitation. As Basque feminist Jule Goikoetxea has pointed out, the bourgeois and patriarchal category of work produces a double effect.[13] On the one hand, it makes visible and positions at the center of the social structure the subject who holds rights and property, thereby reifying activities traditionally associated with or legally limited to masculinity. Despite the deep regime of inequality in class societies, productive activities associated with masculine bodies have found legal protection and economic compensation. This economic recognition had its political correlate, allowing the emergence of what would be known as the subject of rights: commonly able, male, heterosexual, and Caucasian—or what Quandamooka activist and intellectual Aileen Moreton-Robinson calls the *white possessive*.[14] On the other hand, capitalism places activities traditionally associated with women, such as motherhood and care work, in a regime of subordination. As Silvia Federici explained in detail in her book "Caliban and the Witch," the capitalist accumulation necessary to allow the industrial revolution would not have taken place had it not been for the capitalist plunder of unpaid domestic work, slavery, and the conquest and domination of Indigenous populations and territories, which were all contemporaneous events.[15] Federici explains, "When we said that housework is actually work for capital, that although it is unpaid work it contributes to the accumulation of capital, we established something extremely important about the nature of capitalism as a system of production. We established that capitalism is built on an immense amount of unpaid labour, that it is not built exclusively or primarily on contractual relations; that the wage relation hides the unpaid, slave-like nature of so much of the work upon which capital accumulation is premised."[16] Marxist feminism is thus essential to comprehending how digital capitalism operates by rendering work invisible, depriving workers of legal and political protections, and exploiting unpaid affective labor for profit.

However, critical race theory is the body of literature better placed to unpack and make readable the set of oppressive relationships inscribed in the gig economy's labor market. As I explained in chapter 1, a vast number of authors have pointed out how the regime of inequality that defines social and economic relationships is built on historic lines of racialization. For example, Black Marxist and abolitionist intellectuals such as Cedric Robinson, C. L. R. James, and Ruth Wilson Gilmore have explained in detail how capitalist relations of exploitation are built on the Black-white dividing lines emanating from the Atlantic slave trade. This regime of white supremacist domination did not disappear with the abolition of slavery but was maintained through a complex legal, political, and economic framework that privileged the accumulation of white social, political, cultural, and economic capital while socially and physically destroying racialized subjects. Intellectuals from the First Nations of Australia, Canada, and New Zealand have demonstrated how the apocalyptic violence and land dispossession to which they were subjected for the benefit of the white minority were safeguarded by a multitude of (colonizer) treaties, laws, and regulations.[17] These critical voices have exposed settler colonial states as a legal structure designed to shield what Karl Marx defined as original accumulation and more recently Jacky Wang coined as "racialized accumulation by dispossession."[18] Similarly, authors from the Latin American decolonial school have explained with clarity how structures of privilege based on racial caste imposed under the Spanish colonial regime have consistently persisted. Even today, the subjects of privilege in countries such as Colombia, Mexico, Brazil, or Peru continue to be the white minority who disproportionately accumulate economic, political, and social capital in those countries.[19]

This racialized structure of domination persists clearly and visibly in the legal mechanisms that regulate migration, work, social rights such as education or healthcare, and political rights such as voting. The legal and informal scaffolding of Global North countries continues to establish a regime of privilege for traditionally white subjects. For example, it allows free movement of people among the countries of the European Union and guarantees access to the same regime of rights for all citizens of the union independent of where in the EU they choose to live. This regime of recognition of rights does not extend to the citizens of the many European former colonies. For instance, Bolivians in Spain or Algerians in France have to deal with endless bureaucratic obstacles to migrate and regularize their situation or validate their studies. This legalized form of bureaucratic violence is aimed at limiting their job opportunities to the informal and precarious economy, a way of reinforcing the color line delimiting the place they deserve within the system. Similarly, in Australia, the regime of temporary, student, and working visas is purposely designed to produce a subaltern and productive migrant population. Australia establishes numerous prior controls and conditions for visas, generally conditioning access to young, healthy, educated bodies fluent in English whose skills correspond to the jobs demanded by capitalists.[20] The Australian public and private sectors extract wealth from the migrant population through myriad strategies, such as exorbitant study fees, limitations on choosing employers in which is de facto the maintenance of indentured servitude, preventing migrants from accessing the public healthcare system, excluding migrants from the most basic political rights of association and protest, and generally subjecting migrants to a regime of uncertainty regarding their migratory status.[21] This racialized structure is indissociable from the multiple neoliberal processes that have shaken the world. An example of this is the neoliberal regulation of labor relations, which has eroded the structure of labor rights either by extending the retirement age, reducing and controlling access to unemployment benefits, or establishing new labor categories devoid of rights such as holidays, parental leave, or safety. Digital capitalism is happily bridging neoliberal exploitation and racial capitalism to build a digital racialized labor market stripped of rights. The question now is this: how is this carried out?

Digital Factory Lords

Within Marxist debates, a distinction is made between the formal subsumption of capital into labor and the real subsumption. The former refers to the imposition of capitalist extraction techniques on traditional forms of work, namely the salarization of production through mechanisms such as the appropriation of the means of production, the expropriation of land from peasant classes, and the rudimentary extension of the working day. In other words, the process of exploitation changes as domination mechanisms increase in structure, but without a substantial qualitative leap. However, the phase we are experiencing today points towards a phenomenon of a different nature. As various post-Marxist thinkers, including Matteo Pasquinelli and Alessandro Delfanti,[22] have pointed out, neoliberal governmental mechanisms, alongside technological development linked to algorithmic control and surveillance, are allowing capitalists to transition towards what Marx called real subsumption:

a qualitative transformation of the mode of production and the conditions of exploitation of the working class.[23] The new forms of algorithmic surveillance and control are allowing technology corporations to create conditions of labor exploitation that go beyond traditional forms of value extraction in terms of intensity and capillarity. For example, digital exploitation systems quantify and monetize every moment of the workday by imposing an unprecedented regime of surveillance. Regardless of the job performed, such as messaging, transportation, warehouse work, or online teaching, the granularized control and quantification of the worker begins from the moment they log in until they log out. Every meter walked, every delivery accepted, rejected, or made, every word of every message in an online forum, every break for rest and hygiene is recorded in a vigilant registry.[24]

Similar conditions of surveillance would have been considered not only intolerable but a product of science fiction not too long ago. As Mario Tronti reminds us, "It is not the worker who uses the means of production, but the means of production which use the worker."[25] The question to ask is this: in what way does the exploitation that Tronti describes take place, and what defining features does it adopt in digital capitalism? Before delving into the answer, I would like to comment briefly on a quote from Marx:

> There was a constant cry for some invention that might render the capitalist independent of the working man; the spinning machine and power-loom has rendered him independent, it has transferred the motive power of production into his hands. By this the power of the capitalist has been immensely increased. The factory lord has become a penal legislator within his own establishment, inflicting fines at will, frequently for his own aggrandisement. The feudal baron in his dealings with his serfs was bound by traditions and subject to certain definite rules; the factory lord is subject to no controlling agency of any kind.[26]

Amid so much literature and rhetoric, it is sometimes good to remember classic texts like Marx's *Capital*. This specific fragment caught my attention because it brought to light something that often goes unnoticed: the relationship between law, power, technology, and exploitation. After all, the unequal distribution of power and wealth between workers and capitalists stems from the expropriatory ownership of the means of production—energy resources, and land—as well as the capitalist's enormous capacity to predate state-promoted infrastructures. Digital capitalism is not as different as we might believe, as it bases its power on the control of intangible fixed capital, the codes and algorithms of digital machines, and the control of the infrastructures such as cables, roads, and networks that enable the functioning of these systems.

Algorithms, once simple mathematical expressions, have now mutated into digital machines—the critical means of production of digital capitalism. However, companies like DoorDash, Uber, Deliveroo, and Glovo have argued that they are not owners of the means of production in the delivery or transportation industries, but rather that they are simply software companies.[27] In their parallel world, these global companies only provide services to the true entrepreneurs: the riders and the drivers. Digital capitalists claim to be service providers for the true entrepreneurs, their workers! A judge in Spain had to formally state what was obvious to unions, workers, and clients: "The true means of production in this activity are not the bicycle and the cell phone that the delivery person uses, but the digital platform for matching supply and demand owned by the company and outside of which the provision of the service is not feasible."[28]

The digital factory lords not only own the means of production but also the law governing the workplace. One of the distinct characteristics of the new model of algorithmic exploitation is that digital machinery allows capitalists to exert nearly total control over the technical and legal architecture where the labor processes take place. This grants them power over historically contested dimensions of the productive operations such as the working day, wages, and general working conditions. In other words, workers are now confronted with a reality in which machines not only enforce the rules of exploitative production but also embody the essence of labor law. The technological concept of code originates from Roman law, referring to the systematic collection of imperial laws such as the *Codex Theodosianus*. But the similarities between law and computer language go beyond semantic analogies. Both computer code and law are normative languages intended to produce effects—creative, destructive, or regulatory—on reality, whether it is analogical or digital.[29] Both these forms of reality production coexisted peacefully for a while, each with its sphere of influence; the law would regulate the "real" world while code would regulate cyberspace. This undisputed distinction of spheres of sovereignty began to blur with the technological revolution and the rise of the internet, along with the increasing interweaving of the analogue and the digital.[30] The emergence of large digital empires marked the death sentence of the last frontiers between cyberspace and the physical world. For instance, the circulation of knowledge and news, and thus the shaping of political opinion, moved to the digital realm—a territory governed by large digital corporations through the legal code of the twenty-first century.[31]

Computer code has thus taken on a dual position of influence over our materiality. On the one hand, it is the fundamental means of production in the digital era, its factory and machinery, its gears and production plants. On the other hand, it is the determining language of the conditions of exploitation in any work related to the digital realm. However, as a plethora of legal scholars have explained, neither code nor law are neutral or innocent; they are the instruments through which capital inscribes itself into reality. Let us consider a Deliveroo driver transporting and delivering food to some corner of London. This seemingly simple transaction hides the network of traditional laws that are interacting, including labor relations, business relations, traffic regulations, contractual relations with third parties such as bicycle leasing, not to mention the extralegal deals that may occur between the driver and others who—lacking the necessary immigration status—sublease their delivery account. Digital capitalism has added to this already complex scene a dense tangle of new regulations meticulously controlled by digital corporations. Deliveroo reigns supreme over the mediation between the worker, the food-provider, and the end customer. Under the guise of apparent freedom, it regulates which worker will deliver what, how much, and where through a dynamic evaluation and pricing system. Traditional law is somehow suspended in algorithmic dynamics, as it is machines, not human laws, that govern the course of work. In other words, it is a new form of social control based on a cybernetic feedback loop. Obviously, this does not mean that machines have any kind of agency; it is the proprietary code of a corporation whose sole and declared purpose is to accumulate capital. The emergence of this algorithmic exploitation model, where machines are both the means of production and the law of work, is already having serious consequences for the working class. I will analyze some of them.

How Algorithmic Exploitation Augments Racial Capitalism

Let us start with some numbers. In Australia, nine out of ten delivery riders are of foreign origin, with the vast majority coming from Asia (China, India, Pakistan, and Bangladesh).[32] Similar figures can be found in the UK, with a majority of riders coming from the Caribbean and Asia.[33] The rate is also consistent in other regions such as Spain, with most of the workforce coming from Latin America.[34] Yet this dynamic is not limited to the transport and home delivery labor market. According to data provided by Amazon for 2021, just over 65 percent of its hired workforce in its lowest-paying levels were Black or Latinx, and 26 percent were white workers. In contrast, senior leaders were 66.4 percent white, while a meager 10 percent were Black and Latinx individuals.[35] In other words, there is demographic evidence indicating a strong racialization of the workforce in the platform sector. However, the concept of racial capitalism points not only to demographic composition in production relations but to the processes of domination and racialization upon which it is built. In this sense, the above data would not draw as much attention if the working conditions in this sector were equivalent to those of other labor figures. Platform workers earn less, have fewer rights, and in many cases are not even authorized to form unions. According to recent study data, 45 percent of gig workers in Australia claim to earn below the minimum wage, and 69 percent claim to have been deprived of income for being unable to work while sick.[36] For unions, the fact that a huge number of workers have been stripped of basic labor rights has a name: wage theft, and it amounts to more than A$300 per worker per week.[37] This is a huge and legalized transfer of wealth from the poorest working class to the owning classes of digital capitalism.

As Dalia Gebrialhas pointed out, the business model of corporations such as Uber relies on a seemingly endless stock of surplus population—an expendable and replaceable reserve army of workers.[38] Let us consider the model of companies like Uber. What does their advertising promise? Their commercials offer us the dream of the capitalist market—total access to unlimited goods and services, "Get almost everything"[39]—without being restricted by annoying regulations or schedules; "Doors are always opening."[40] Digital capitalism markets self-indulgence, the promise of "Do less," the fantasy of power over a technologically enhanced servitude.[41] But what does their business model depend on? How does it impact workers? To maintain that control over workers, Uber relies not on slavery or serfdom, but on the violence with which the state and capital coerce a precarized population, producing them as racialized subaltern subjects. In other words, the condition of existence of digital capitalism rests on the mode of exploitation of the racial, which becomes evident at three levels: the legal framework that, by violating labor regulations, produces workers as entrepreneurs; the technolegal tools that operationalize and regulate the inflow of surplus population, thus ensuring total and immediate availability of labor; and the algorithmic, vigilant, and punitive control over productive processes.

Digital capitalists have allocated a huge amount of resources to legalize a model of labor exploitation that clashes with the traditional distinction between employer and employee by seeking to classify gig workers as independent contractors. The goal is clear: these corporations seek to monopolize all the power that capital confers on their workers without having to fulfill any social or economic responsibilities, such as unemployment insurance, health care, and family leave.[42] This has been met with resistance from workers, who have sought to fight this offensive against labor rights. However, despite some timid victories, capital is demonstrating its enormous capacity and power to impose regimes of exploitation in clear collision with social needs, as evidenced by the California case. Following a broad mobilization by progressive sectors, the state of California managed to pass legislation in 2019 that, beyond capitalist legal engineering, obliged corporations to recognize workers for what they were: workers. Digital capitalists were not willing to accept the rules of the game, so they immediately unleashed a costly lawfare and lobbying campaign (over $200 million in propaganda). They eventually succeeded with Proposition 22, thus creating an exception to the regulation that left digital platform work unprotected.[43] The judiciary itself considered this to "limit the power of a future legislature to define app-based drivers as workers subject to workers' compensation law and is therefore unconstitutional."[44]

It is important to note that although digital corporations prefer and try to obtain legal support for the creation of this worker-subject stripped of rights, they do not hesitate to operate illegally. This is what has happened in countless jurisdictions, including the United Kingdom, Spain, and France, where, despite the existence of specific rulings and legislation obliging corporations to recognize their workers as employees, they continue with their antiworker practices.[45] As explained previously, the digital lords maintain this relative autonomy with respect to the legal infrastructure of the state thanks to the control they exercise over the means of production. It is the digital platform—as a space of production and labor law—that articulates and modulates the army of disposable surplus, forcing them into absolute availability and surveillance against the interests and health of the worker.

The digital lords push workers to be in a permanent state of alertness—in other words, constant connection. The more time they dedicate to the platform, the more possibilities the worker has to access clients or better slots in work schedules. For the majority of this connection time, availability is not paid, despite the fact that this absolute availability is sold to consumers by the platforms as their most valuable asset.[46] It is necessary to pause here to explain one of the most significant strategies employed by digital corporations: payment per piece. Many of the labor struggles and conquests of the nineteenth and twentieth centuries pursued the formalization and salary of a well-defined and delimited schedule regime. This was done to guarantee a minimum income and work hours, a mandatory rest regime, and a series of measures aimed at protecting the health of the workforce. However, as researchers such as Veena Dubal point out, "innovative" digital industries in sectors such as taxis and home delivery have managed to universalize a form of remuneration per piece instead of per time worked.[47] Workers have no control over the mechanisms that regulate the price of the services they provide, supposedly established through dynamic processes generated by each corporation's proprietary algorithmic software. Marx already pointed out the advantages that the payment per piece had for capitalists: "The quality of the labour is here controlled by the work itself, which must be of average perfection if the piece-price is to be paid in full. Piece wages become, from this point of view, the most fruitful source of reductions of wages and capitalistic cheating."[48]

This triumph rests on previous neoliberal attacks which, as Dubal demonstrates in her masterful historical account of the taxi sector, have been eroding worker's rights for decades.[49] In other words, through piece wages and algorithmic subjection to availability, corporations revive old fantasies of racialized servitude.[50] The viability of this model rests on a premise that is not always possible to fulfill—that the demand for services greatly exceeds the supply of work. How does the corporation maintain control over its *independent* workers? How does it manage to maintain standardization over its processes? Furthermore, what happens when the supply of work exceeds demand? Corporations use various complex mechanisms to keep their workforce disciplined, of which I will highlight three: monopolizing information, mediating with customers, and rating workers.

Digital corporations enjoy an asymmetrical informational power over their workers, which puts them in a position of dominance and power over the production processes. This informational power is exercised before, during, and after the work is performed. For example, corporations like Uber know the personal, biometric, economic, and legal data of both their workers and their vehicles. Workers have to give up their right to privacy, at least while they are connected.[51] Similarly, the corporation controls the identity, credit cards, and devices of its customers. When offering a ride or delivery, drivers and riders are unaware of fundamental elements of the work they will perform, including the final destination of their client or merchandise and, consequently, the route they will take. During the ride, Uber puts into practice what Rabih Jamil has called the "algopticon," an algorithmic surveillance regime that uses the driver's and customer's smartphone to monitor a wide variety of data, ranging from driving habits or the route taken to customer feedback.[52] This not only allows the corporation to accumulate a vast volume of information but also to threaten and discipline the driver in real time, through "recommendations" and "advice." The informational asymmetry continues after the work is done, as corporations maintain the monopoly of the link with customers, which—as work dynamics demonstrate—belong to the platform rather than the driver, as it proclaims.[53]

Theoretically, the way platforms facilitate the meeting between the supply and demand of services responds to a series of "technically neutral" criteria algorithmically managed, including availability, distance to the client, and connection time. However, as workers know well, one of the fundamental elements that define the digital corporation's control over production processes has to do with the rating of the workers. This operates at two levels: the rating of the workers by the company, and the rating of the workers by the client. The way algorithms assign values to workers is not public; it is a trade secret protected zealously by corporations. The workers' ratings can be a numerical or symbolic value (for example, seven out of a maximum score of ten or four stars out of five). The higher the value, the more likely a worker will be assigned a task, and if it falls below a certain threshold, workers are disconnected—the digital capitalist name for termination.[54] In this way, the client and the corporation become disciplinary agents, monitoring in real time the performance of operations. The worker, often a racialized subject, is pushed by both the client's expectations of servitude and the power of the rating to overcompensate affectively (smiles, good treatment) and materially (water, sweets) for the clients,[55] which workers have repeatedly denounced as denigrating and stressful.[56]

The algorithmic evaluation regime fosters hypercompetitiveness among workers through rewards (bonuses), gamification, promises (reward zones), and punishments (penalization for rejecting a delivery threshold). They are forced to blindly accept rides and deliveries not knowing if the journey will be economically profitable for them and to work during low-demand hours or under conditions that they would not normally perform their work (for example, having to accept dangerous routes or work under extreme weather conditions). This not only entails an economic disadvantage for the workers, who see how the corporation steals a part of their salary, but it also puts their lives at risk. For some, this is a symptom of the necropolitical regime imposed by digital capitalism, which—in total disregard for the health of its workers—compels them to work in dangerous conditions, while at the same time shirking their responsibility when they are injured or die.[57] These unpaid wages and this erosion of their bodies are the price that workers have to pay to obtain a high rating that allows them to access compensation in many cases below the minimum wage.

Fight Back

"You don't work for us, you work with us. You don't drive for us, you perform services. There are no employment contracts, no performance targets, only delivery standards. You don't receive wages, but fees. You don't clock in, you just become available," said Gavin Maloney, a manager at the fictional Amazon-like corporation "PDF," to Ricky Turner. Ricky, middle-aged and desperately unemployed, was ready to join the ranks of exploited entrepreneurs (read, workers), as a delivery driver in Ken Loach's 2019 *Sorry We Missed You*. The film portrays the everyday drama of precarity and the impossibility of making a decent life in a dehumanizing capitalist system. However, as usual, reality is much worse. "I shouldn't have to work 60 hours a week just to pay bills," declared a UK Amazon worker to the *Guardian*, and he was right. The interview continued to describe what it is like to work under constant surveillance, overworked and underpaid even (especially) during the pandemic.[58] A pregnant worker spoke in a similar tone, saying, "I am pregnant and made to stand and work for 10 hours without a chair. They constantly mention my idle time and tell me to work hard, even though they know I am pregnant. I feel depressed when at work."[59]

Workers in the platform sector are paid less for their labor and have limited access to social rights that are generally available in other sectors, such as healthcare, vacations, and leisure time. They also face more intense and extensive surveillance at work, which not only questions their rights as workers but also their dignity as people. Furthermore, as I have shown, the regime of racial technocapitalism has necropolitical overtones, with workers exposed to a regime of risk and exposure that is neither recognized nor assumed by corporations. Despite being technologically induced, the real subsumption is interdependent on the legal, political, economic, and cultural infrastructure of the racial capitalism described above. In other words, it can be argued that algorithmic technologies amplify the regime of precariousness, flexibility, and temporality of racial capitalism through the evaluation regime with which platforms control both the worker's availability and the assignment of tasks to them. What can be done in such a hostile scenario?

In 2021, the world bore witness to one of the most prominent unionization campaigns, which took place at Amazon's warehouse located in

Bessemer, Alabama. Despite facing initial obstacles, this movement managed to garner substantial attention and support, shining a spotlight on the mounting dissatisfaction among employees regarding their working conditions, surveillance methods, and the relentless pace of their work. This grassroots initiative acted as a catalyst, igniting similar drives for unionization in various other Amazon facilities throughout the United States and even beyond its borders. The initial setback in Bessemer didn't deter the movement; instead, it triggered a nationwide surge in activism and heightened awareness concerning labor rights within Amazon and the broader tech industry. This was something they achieved despite the mafia like tactics deployed by Amazon: "Amazon really instills fear in workers. It wasn't just that there were anti-union posters everywhere; Amazon hired a ton of union busters that were constantly walking around the building talking to workers. It was intimidating."[60]

Furthermore, *Jacobin*, in a 2022 article, celebrated the groundbreaking victory of Amazon workers in Staten Island, New York. These workers successfully formed the very first US union within the company, marking a historic milestone in the labor movement within the tech sector. In the words of one of the union organizers,

> This is still surreal for me. I can't believe that two years to the day I was fired, JFK8 made history on March 30, when the voting ended. To see over five thousand people vote in person, to go in and out of the tents, to sign off on the sealed ballots, to witness the ballot count live and in person at the NLRB office—I can't put that into words. That experience is something that I will never forget for the rest of my life. I said to my team: Win, lose, or draw, we already made history. I want to let everybody know that I never had a doubt in my mind that we would win. We were going to win. Even the last observers that I had in the tent—the last signature that I signed on the sealed ballot was "ALU for the win." I wanted them to know that we knew and believed that we were going to win the election.[61]

UK Amazon workers have also been denouncing the oppressive working conditions and the harsh system designed to exploit the fragile and increasing surplus of disposable workers. They have been speaking out—but no one was listening, so they organized. Just one month after the interview just quoted was published in the *Guardian*, in December 2022, the

UK witnessed its first strike by Amazon warehouse workers. In just four months, there have been fourteen days of protests, pickets, and strikes, where among other slogans could be heard "We are not robots" and "To unionize is not a crime."[62] Organizing this protest was not easy. Amazon is known worldwide for its antiworker practices, the persecution of union leaders, and its brutal efforts to crush any attempt at organization in its fulfillment centers. But national and international solidarity prevailed. Learning from the recent successes (New York) and failures (Alabama) of the movement in US warehouses, UK Amazon workers are articulating a strong movement. Their first demand may seem anachronic, but it is a symptom of the times we live in: that their union be recognized.

In Catalonia, I had the privilege of meeting Nuria Soto. She was one of the first riders for companies like Glovo and Deliveroo. She was also one of the first to rebel against the exploitation and mafia tactics of capitalists. In 2017, along with other riders, she founded one of the world's first unions in this sector: Riders x Derechos. Despite the power asymmetry, the riders managed to overcome exploiters, expand networks of solidarity with other labor organizations, and organize the first strikes in the industry. After an intelligent negotiation process with the state, the riders won the first law (2021) declaring their status as workers. This has certainly not been the end of their struggle. Most companies in the sector have continued to operate in Spain illegally, implementing sophisticated strategies to evade their responsibilities as employers and multiplying their antiunion efforts. In this scenario, groups of organized workers have decided to form delivery cooperatives under worker control such as La Pájara in Madrid or Mensakas in Barcelona. The latter operates under the open-source platform created by CoopCycle, a federation of bike delivery co-ops. As can be seen, the response to the exploitation of digital capitalism can take many forms, but they all come from a common place: organization and resistance. I think that there is no better end to this chapter than the words of Nuria Soto: "We are becoming more and more relevant and that is what makes us stronger and stronger. We are what the companies don't want to exist. That is to say, workers saying that we don't want this, that we are not our own bosses. In a way, we are the failure for the companies and we have to continue to be that. As long as we are the failure of the companies, i.e. the opposite of what they want, we are on the right track."[63]

5

Law and Extractivism

Without a secure supply of critical raw materials, the Union will not be able to meet its objective for a green and digital future.
—European Union Critical Raw Materials Act 2023

I don't know how many times I've played it, but every time I click play on the promotional video of the lithium mining project that Lithium Iberia is promoting in Cañaveral (Extremadura, Spain) the effect is the same: first nausea, then rage.[1] Before the end of the first minute, almost all the key concepts of "green and digital" capitalism have already made their appearance: sustainability, environment, the European Union, green and circular economy, clean energy, new mobility, lithium, and, my "favorite," Green Mining 4.0. A voice-over recites the virtues of the project while images of idyllic landscapes, electric cars, and the fast and wild flow of a free river (which river will this be in the Extremadura of swamps?) play one after the other. The video lauds the Swiss landscapes and German industry, all thanks to mining that is "especially respectful of environmental, social and cultural constraints."[2]

According to corporate publicity, during the approximately twenty to thirty years that the exploitation will last, four hundred direct jobs will be created, a thousand indirect jobs, and another three hundred in the cathode factory that a nanotech company of dubious transparency, Phi4tech, has projected in the same locality. That's saying a lot in a region particularly affected by depopulation and unemployment (in 1900, there was twice the population that there is now). According to Lithium Iberia's statements, the thirty thousand tons of lithium per year will contribute to the production of 2.5 million electric car batteries, facilitating not only the transition towards intelligent and environmentally committed transport but also supporting one of the European Union's flagship projects, while boosting the value chain in the depressed region.³ There is no other solution: green and digital capitalism or a world like that of *Mad Max*.

But no matter how much I watch the video, I cannot recognize the territory through which I walked with the activists of the No to the Cañaveral Mine Platform. The crude animation cancels out the complexity of the *dehesa*, the web of life that jumps from the cork oaks to the holm oaks, which extends from the nearby mountain ranges through the surface and the subsoil of the land that green and digital capitalism wants to exploit. Perhaps that is why the propaganda forgets to mention that the exploitation is proposed in what were historically the communal territories of the region, hectares of forest that escaped privatization and logging thanks to its traditional, plural, and collective use by the inhabitants of the villages among which it is distributed: Grimaldo, Holguera, Torrejoncillo, Casas de Millán, Pedroso de Acím, and Cañaveral.⁴ All of these villages have an immense historical and social heritage. It is not a forest; it is a way of life.

The corporate science fiction film avoids mentioning that the extractivist project will affect Red Natura 2000 and the Zonas de Especial Protección para las Aves (Protected Birdlife Reserve under European and Spanish laws) areas such as Monfragüe (European areas considered to be of in need of special protection), which are fundamental for the reproduction of endangered species. It also omits the capitalist gluttony for water and the water stress to which the territory will be subjected. Lithium Iberia has requested to extract sixty-nine liters of water per second from the Galisteno aquifer, which is vital for the ecosystems and communities of the area.⁵ They request this in a context of global warming and chronification of the drought already endemic in a large part of the Iberian Peninsula, at a time when "the drought," the disease that has already spread over thousands of hectares of holm oaks and cork oaks, silently threatens the trees weakened by the lack of water and humidity.⁶

The video forgets the explosions, the sludge, the slag heaps, and the clouds of waste that will float down to the water reserves from which the people of Cáceres drink. The video does not mention the thousands of euros used to divide the people, to buy loyalties in impoverished areas, capitalist crumbs thrown here and there with the characteristic disdain of the powerful. Sustainable digital capitalism, the solution to all global problems, nevertheless requires the sacrifice of areas that have been giving everything—resources, people—for centuries. It demands they be exploited, emptied, and cornered. I ponder this as I finish watching the video, go back to the field notes, and review the *greenwashing* articles bought by the company in the local papers. And yes, today there is a basketball team named Sacred Heart Lithium Iberia.[7]

This is not an isolated initiative. The Extremadura lithium hub would be accompanied by another contested mine two kilometers from the urban center of Cáceres. According to statements by its promoters, the San José (Cáceres) lithium mining project could host the second-largest hard rock lithium reserve in the EU. This project, they say, is vital for the region and for the sustainable and green development of Europe. It seems incredible that anyone would consider it viable to develop a mining project two thousand meters from a population of more than ninety-six thousand inhabitants. And Cáceres is not just any city. Its alleys are saturated with history, crossing its walls is like jumping into a universe where you can breathe the complex memory of a past superimposed on infinite and imbricated layers. You don't walk there, you navigate a complex geology of history. The Jewish and Muslim suburbs surrounding the medieval wall give way to the palaces built thanks to the conquest of the Americas, American dispossession, and plundering (a visit to the palace that the hispanized Aztec nobility of the Toledo-Moctezuma family built in the historic center is a must). Strolling through the city's labyrinthine streets, traveling up to the wide bourgeois avenues, is like walking through history. Perhaps you have even seen it. When I was doing fieldwork in the city, I came across the surprise that a good part of its streets were cut off due to the filming of *Game of Thrones*. Who would have thought that *"King's Landing,"* a world heritage city, would be threatened by a mine destined to satisfy the needs of the ecological cars and computers of the green energy transition.

The San José project is promoted by a financial and business network operated from Australia. Infinity Lithium is the corporate face of this alliance of extractive capitals, operating locally under different and changing formulas as the subsidiary company Extremadura New Energies. The project has received approval from both regional Spanish and European authorities and was the first lithium mining company to receive €800,000 in green funding from the European Commission's Strategic Action Plan on Batteries (2018). Moreover, at the end of 2023, the Spanish

Ministry of Industry confirmed a grant to the mining company of nearly €19 million to carry out an "integral lithium treatment project." Paradoxically, this transfer of Spanish public funds to Australian capitalists takes place within the framework of the Recovery, Transformation and Resilience Plan financed with Next Generation Europe funds. In other words, money earmarked for Europe's green transformation is being channeled to mining its impoverished Southern European peripheries.

Anyone looking at Infinity Lithium's website might think they are a charity or even an environmental organization. According to them, "At Infinity Lithium our focus is to help facilitate Europe's energy transition through the development of our fully integrated San José lithium project, and the development of innovative, sustainable lithium processing technologies through our Infinity GreenTech business."[8]

The obscure company has been knocking on the doors of the powerful and buying corporate and journalistic souls for years in order to smooth over the exploitation of one of the most naturally, socially, and symbolically important spaces in Cáceres. It is difficult to convey the affinity and link of the region with Sierra de la Mosca, or La Montaña (the Mountain), where the lithium mine is projected, but I will try to do so. Imagine a mining operation in the middle of Collserola (Barcelona), Hyde Park (London), or Maungawhau (Auckland-Tāmaki Makaurau). Imagine that it is not only a territory with an intrinsic natural and emotional value but that the mine is also built on the aquifer that historically nourished the population. Add to that the symbolic value of being the place inhabited by La Virgen de la Montaña, patron saint of the city, a place of pilgrimage and devotion, and a place from which every year the image of the Virgin descends for tens of thousands of cacereños to honor her. Every time I have asked someone what the Mountain was for them, regardless of their feelings toward the mine, toward politics, they have answered me in a very similar way: it is ours.

This sentiment explains the initial (and progressively eroded) frontal rejection of the mine by the city of Cáceres. Politicians, even those of the Spanish Socialist Workers Party (formerly in government), who were so comfortable professing their extractivist, developmentalist, and plundering faith, could only follow the citizens in their almost unanimous rejection of the project. It was not a matter of heart, nor of beliefs, but just a simple exercise of political survivalism. But Cáceres is tired. It is tired of seeing young people leave, losing opportunities, being on the wrong side of history. It is afraid of disappearing, becoming even poorer, sinking, perhaps hopelessly, into precariousness. And, as the local elections have shown, fear and uncertainty are more powerful than affection and the defense of the common good. It does not matter that, as the geologist Juan Gil told me (and multiple judicial instances have confirmed), the project is nonsense. An attack against the aquifer of El Calerizo, against protected species, against the existing and potential sustainable productive fabric of the Sierra de la Mosca. For the geologist, the situation of the megaproject in the middle of the Sierra de la Mosca is only justified by spurious reasons, ease of connection and access, and economic savings—in short, the pernicious neoliberal efficiency of which we already know the devastating environmental and social cost regardless of reason, science, or sentiment. The exhaustion of a significant part of the people is so brutal that some are willing to sacrifice one of the most significant expressions of their identity for a meager chance of a future, even if it is a lie.

There are, of course, those who resist. The Plataforma Salvemos la Montaña (Let's Save the Mountain Platform) has been organizing and fighting since 2017, raising resistance in the streets and in the courts. Together with other organizations such as Ecologistas en Acción (Ecologists in Action), they managed to stop the initial plan for an open-pit megamine proposed by Australian capital (the new one will be "sustainable" and subterranean). The social mobilization managed to paralyze the permits and actions, and, most importantly, they demonstrated the permanent illegality in which corporations such as Infinity Lithium live and prosper. But it is not enough. Capital is patient and has resources (private and public) that overrun those that can mobilize activism. This situation has been contributed to by public figures and institutions accustomed to doing shady deals, trafficking with their land, and surrendering themselves into the arms of internal energy colonialism.[9] Proof of this is the meeting held by the front men of the Australian extractivist capital with the former president of the Junta de Extremadura.[10] Proof of this is also the strong support for the mine from the conservative president of Extremadura who came to power in 2023 with the support of the extreme right who cleverly weaponized the "jobs versus environment" false dilemma.

Massive resource extraction is presented to the population not as something probable, but as something inevitable. The old discourse of the historical inertia of progress and civilization is now tinged with the benevolent aura of sustainability and the Promethean fire of digitalization. But that is only a facade. The logic of extractivism, a historical regime of capitalist accumulation and dispossession based on the appropriation of natural resources and the commodification of people and land for the benefit of investors, is the real engine of this project, as it is of those proposed by Chinese, Australian, Canadian, and American capitalists in Jujuy (Argentina), Atacama (Chile), Jadar (Serbia), and Barroso (Portugal). In each of these regions, green extractivism is imposed following logics of structural, historical, economic, and physical violence, as diverse in their methods as they are in their objectives.[11]

In Jujuy (Argentina), which has one of the world's largest lithium reserves, the regional government changed local laws in 2023 to allow the advance of mining while facilitating Indigenous dispossession (and repression).[12] A few months later, the controversial and *libertarian* Argentine president Javier Milei announced, to the applause of Elon Musk (CEO of Tesla), a new wave of mega-mining and extractivism in Argentina. Musk is perhaps one of the most notorious examples of the power and scope of global digital capitalism and its close relationship with green extractivism. Musk not only supported Milei in Argentina's controversial electoral campaign, he has already expressed his support for the so-called lithium coup that ousted the leftist and indigenist president Evo Morales in Bolivia. Extractivist violence takes on a more subtle and patient face in the peripheries of southern Europe (Portugal [Barroso], Spain [Extremadura], and Serbia [Jadar]) than in the former colonies. Less brutal in means, but with the same objectives, governments and corporations launch campaigns of social engineering and counterinsurgency, criminalizing environmental movements branded as radical, creating divisions through the unequal distribution of aid and subsidies, and putting into practice large-scale political bullying.[13]

According to state and corporate propaganda campaigns, Extremadura, together with Barroso in Portugal, would go from being the poor periphery of southern Europe to being at the head of the new green and digital industrial model to which Europe wants to move. That is why red carpets continue to be rolled out for inner colonialism, for green

Figure 5.1: Milei 2023.

capitalism. Phi4tech has already begun construction in Badajoz of "the first battery cell factory in southern Europe," the spearhead of an ambitious integrated energy storage project. The regional government applauded the statements with jubilation, accompanying it with legislation tailored to green extractivism, the so-called Extremadura lithium decree (2022). This legislation, in line with the new Critical Raw Materials Act (2023), smooths out the already tepid administrative and environmental procedures for mining in the region. To paraphrase Tlingit activist and academic Anne Spice, governments mobilize the language of critical infrastructure to transform industrial-capitalist lithium or copper mining projects (among others) into crucial projects for general interest.[14]

The narrative is clear. Those at the top need Extremadura's resources for the new era of wealth and development. In exchange for the sacrifice of some landscapes typical of the failed economy of the past, green and digital capitalism will bring progress, jobs, and the promise of protecting social and natural life in a territory hit by poverty, unemployment, and the forced expulsion of population from small cities and towns. There is no choice, the train of history will pass—what we have to see is if it will run us over or if we can get something out of it. This last remark is one I heard from a shoemaker in a working-class neighborhood of Cáceres, who said them loud and clear while another customer nodded in agreement. It is certainly not all doom and gloom and melancholy acceptance. The self-organized population in places like Jadar has rejected, through mobilizations, blockades, direct actions, and training activities, the attempts to impose mining megaprojects.[15]

In the following pages I will attempt to dismantle the discourses that present the digital economy as harmless, green, and free of material consequences. This narrative equates success with economic growth and thus perpetuates a dangerous paradigm that threatens to initiate another cycle of socio-environmental devastation. Given the breadth and relevance, the analysis will be divided into two chapters that will attempt to answer the following questions: How have the institutions of digital capitalism managed to disguise the catastrophic environmental damage they have caused as necessary and socially responsible practices? What kind of impacts are being caused by the routine extractive operations imposed by digital capitalism? Do we have the discursive, legal, social, and political mechanisms in place to address these impacts?

The Imperial Infrastructural Matrix

In recent years, a refreshing new wave of thinkers has emerged from the adjacent fields of decolonization, activism, anti-capitalism, internationalism and abolitionism inviting us to question the materiality and genealogies of dispossession underlying what they have come to call the infrastructures of empire and death.[16] Authors such as Anne Spice highlight the political and colonial character of the presumably neutral concept of infrastructure, which, following Brian Larkin, understands as a complex assemblage upon which social life is built. Larkin writes, "Infrastructures are built networks that facilitate the flow of goods, people, or ideas and allow for their exchange over space. As physical forms they shape the nature of a network, the speed and direction of its movement, its temporalities, and its vulnerability to breakdown. They comprise the architecture for circulation, literally providing the undergirding of modern societies, and they generate the ambient environment of everyday life."[17]

Life, yes—but what life? For authors such as Spice, Shiri Pasternak, and Leanne Betasamosake Simpson, the forms of life that emerge from the infrastructural assemblages of contemporary racial and colonial capitalism are incompatible with any other form of social or natural life that is not subsumed in the destructive logics of capital. Spice sharply questions the discursive logic of the Canadian colonial state in its qualification of highways, gas pipelines, and energy networks as critical infrastructures. She redefines them as forms of colonial violence—invasive infrastructures—that normalize the destruction of rivers, pastures, and forests in the name of the "general interest" while impeding, limiting, and brutalizing the infrastructures of peasant and Indigenous life. For her part, Deborah Cowen excavates the infrastructures of empire as a way to understand contemporary racialized settler colonial scenarios.[18] Both Spice and Cowen extend our understanding of the infrastructural not only to the socio-technical assemblage but also to the legal instruments that enable the reinscription of white colonialism over Indigenous and peasant jurisdictions, sovereignties, and territories. In doing so, they link into the fundamental critique of colonial law as a fundamental element of white-capitalist-patriarchal dominance raised by Irene Watson and Aileen Moreton-Robinson among others.[19] In a brilliant exercise of reappropriation, Spice proposes to think the notion of infrastructure outside the logic of capital, of domination: "If we think of a river as infrastructure, then it's not something that is built and then walked away from, nor something that just exists in space as material. It may have that capacity, but it's also something that requires constant maintenance and care. It's something within which we are also embedded—a web of relations—maintaining and holding up Indigenous and natural law, making sure that our rivers stay healthy. If our attention is drawn to those healthy relations, then we are going to treat these infrastructures differently."[20] This is what, supported by activists and intellectuals Fred Moten and Stefano Harney, Spice defines as fugitive infrastructures. But as we know, these fugitive infrastructures are persecuted, under siege. They live under the permanent threat of their destruction and the criminalization, if not the assassination, of their defenders.

Despite their omnipresence and hegemony, the infrastructures of death remain hidden from our gaze, simply because they are fully normalized. The social, environmental, and moral destruction that comes with their mere existence is inscribed in the normal course of everyday events, invisibilized under an ideological cloak that blocks our possibility of reaction. It is a state of permanent domination that has as much to do with the violence with which the alternatives are persecuted as it does with the addiction that their "comforts" generate in us. Of course, we must not forget either the immense work of the ideological apparatuses in the reaffirmation and banalization of a way of life based on horror, which is a pedagogical function of the oppressor that educates us in distance, insensitivity, and the renunciation of responsibility for our individual and collective acts. It is a pedagogy of domination that distances us from the possibility not only of constructing but also of imagining alternatives. This framework, which is as epistemic as it is economic, political, and social, is not watertight. It leaves escape valves that allow the (contained) escape from horror in the form of a mirage of collective agency: regulation. The European Union is a specialist in these types of collective fantasies, in this territory of capital there is a proliferation of maximalist legislations aimed at "protecting citizens from the power of big tech," at defending their privacy, at promoting ethical artificial intelligence, if not the general framework of human rights. However, this pale illusion contrasts with a reality defined by the domination of a handful of capitalist, technological, financial, energy, or arms companies that govern the prevailing extractive, punitive, developmentalist, colonialist, and racist logics. In the face of the prevailing narrative, it is necessary to affirm that digital capitalism is not innocuous. The cloud is heavy.

Every click, every message, every post, every audio file, every song we listen to, every video we watch, and every document we create online has an echo, a resonance in the material world. Whatever ethereal metaphors such as "the cloud" may indicate, the Internet has weight, volume, guts, and body. The digital does not exist apart from the physical reality in which we live; the network is as physical as it is digital. There is no streaming without cables, no online shopping without production, machinery, data centers, and vast logistics chains. There are no platforms without power. The huge infrastructures on which digital capitalism is based are connected to extractive, industrial, and logistical processes that could not exist outside a globalized capitalist economy. Think, for example, of the smartphone that almost everyone carries with them and which, according to various studies, is the main gateway to the Internet. Any smartphone contains, among others, boron, palladium, tungsten, carbon, silicon, cassiterite, gold, aluminum, lithium, tantalum, copper, zinc, beryllium, indium, nickel, cobalt, silver, neodymium, europium, and terbium. Each of its elements requires extraction, processing, and manufacturing in devastating processes that turn the element into raw material in factories located thousands of kilometers from both the mining sites and the final consumers.[21]

Let's focus on the instrumental element to materialize an important part of the "digital and sustainable economy," which I will explain in more detail in the following sections. Silicon is not a rare material; it can be found in almost every corner of the Earth. It is likely that during a hike, you have passed over its beautiful, unprocessed form: quartz. Once mined, quartz must be ground into fine sand. The sand is then subjected to an electrochemical melting process at a temperature of 1250–1350°C degrees, resulting in silicate metal. However, to produce electronic components such as microchips, polysilicon is needed, a highly purified form of the metal (99.99999999999 percent) that is obtained by complex chemical procedures and a sophisticated melting process that can only be carried out using high-purity quartz crucibles.[22] The problem is that the required ultra-high-purity raw material is produced and controlled on a near-monopoly basis worldwide by the Belgian mining company Sibelco, which boasts of producing the purest quartz in the world at its Spruce Pine mine (North Carolina).[23] The silicate crystals are cut, polished, and distributed to microchip manufacturers, among others. Multiply this process by each of the elements described above, from gold to tungsten, and this is just the beginning of the smartphone supply chain.

Digital capitalism exploits the unequal global distribution of wealth, power, and control over capital flows by taking advantage of the lax labor and ecological regulatory frameworks of the Global South. The lax labor and regulatory frameworks which serve to legitimate capital accumulation (including primitive accumulation) are constructed carefully through processes of neocolonial and imperialism, establishing what David Whyte defines as "regimes of permission."[24] Thus, once the raw materials are obtained and processed, for example, in the Andes, they are distributed to industrial megacities (often in Asia) where the various components of the smartphone, including lithium-ion batteries, will be manufactured in heavily automated factories.

A case in point is the Foxconn plant in the city of Zhengzhou in China, where an army of three hundred thousand workers who live on site in dormitories has already produced more than 250 million phones, becoming the largest manufacturing site of Apple's iPhone.[25] Pollution levels in these enclaves are extraordinarily high, leading to a plague of respiratory problems. Even China's lenient pollution standards highlight cities like Zhengzhou as some of the worst places to live. Once manufactured, the various parts of the smartphone will then be distributed to assembly and packaging factories. From there, they will depart in trucks for huge mega-ports, where cranes more than one hundred meters high will stack thousands of containers on colossal cargo ships, such as the Ever Alot which flies under the flag of Panama. At 399 meters long and 62 meters wide, the Ever Alot is capable of transporting twenty-four thousand containers (about 240,000 tons) anywhere on the globe.[26] Shipping is itself one of the most polluting industries on the planet, responsible for 3 percent of greenhouse gas emissions.[27] After receipt at the port, the smartphones will be redistributed (mainly by truck) to different logistics centers, from where they will be shipped using digital tech to retailers and consumers who have most likely relied on online shopping. There is no environmentally friendly alternative to this supply chain, which integrates Andean lithium, North American quartz, African platinum and gold, and Middle Eastern oil, along with Asian rare raw materials.

Regardless of their condition, smartphones are used for an average of twenty-four months, after which they are replaced by a new model. Despite their sustainable and green promises, corporations like Apple owe their loyalty first and foremost to their investors and shareholders, who expect infinite sales growth, as the company promises every year in their letters to shareholders.[28] In the short period between April 2020 and September 2022, Apple launched eight different phone models and consequent marketing campaigns, thus fueling unsustainable consumerism. Moreover, this company has admitted that (mandatory) software updates to its devices have had a negative impact on device performance, in a clear example of planned, engineered obsolescence.[29] Although some of the elements of the phones will be recycled, this process carries its own environmental impacts. Even considering the possibility of an increasing reliance on recycled materials, which for items such as lithium is simply not possible, the results would never be sufficient to meet current demand, let alone sustain the expected and much desired growth. As much as corporate greenwashing talks about "circular economy" and "recycling," the truth is that each new generation of smartphones reproduces the extractive-productive-logistical scheme summarized above.

However, a smartphone does not work in isolation; it requires a connection. The old landline phones were relatively simple. They relied on a network of copper wires that, suspended on poles (made from some 130 million trees in the United States alone), connected the exchanges to the users. Today's Internet connections require many more elements. The most visible and elemental are communications towers, which can range in size from small urban devices to huge 4G and 5G towers. In the United States alone, there are some 418,000 cell sites.[30] Each of the hundreds of thousands of towers that dot the landscape is composed of a concrete base, a galvanized iron tower, and receiving devices. This structure must also be connected to the power grid, and the largest antennas typically have a diesel-dependent backup power system that is in permanent use.[31] But what are we connecting to?

If we want to access a web application or a digital platform located in the swarm known as the Internet, this requires the existence of millions of computers, routers, and cables connected to data centers. Data centers can range from relatively simple devices made up—like cell phones—of hundreds of elements to vast mega-machines such as those located in the Virginia cluster or those in Oregon and Washington where giants like Amazon are located.[32] For example, the 120,000 square meters of Google's data center in Denmark (an investment of €600 million) integrate solar farms, which are still insufficient to meet the energy demanded by the mega-machine.[33] Other data centers do not boast allegedly clean energy. For example, the new centers planned by Amazon in Oregon will run on natural gas from Canadian fracking.[34] This same energy source also powers the aforementioned Virginia cluster, where energy gluttony has been demanding the creation of controversial new pipelines for years. But energy is not the only thing these centers devour. A center like Google's in Arizona uses between 3 and 15 million liters of water per day, an astronomical amount in a territory characterized by permanent water stress. Another Google data center in South Caroline has free access to pump 1.9 millon liters of water per day.[35] The combined water withdrawal of Google, Amazon, and Meta was an estimated 2.2 billion cubic meters in 2022.[36] Data centers in the United States consume at least 1.7 billion liters of drinking water per day.[37]

These centers are in turn connected to the network and to centers on other continents through submarine cables made of a special alloy of iron, copper, and fiber optics. An example of this is the Curie, one of Google's six private cables, which—over 10,500 kilometers—connects the United States with Chile and enables the 72 Terabytes per second of data needed for services such as YouTube and Gmail in South America. However, this is only a glimpse of the environmental impact required to access the Internet via smartphones.[38] The material operations necessary for the functioning of digital capitalism, even for something as mundane as uploading content to the cloud from a smartphone, carry a profound and devastating ecological impact. The state and corporate actors responsible for these effects often escape legal or political consequences, benefiting from the culture of impunity in which they operate. Not only do these significant environmental harms go unpunished, but, as I show in the following sections, they are also dishonestly presented as steps towards a greener, more sustainable economy.

In recent years, greenwashing campaigns have proliferated in the big tech industry. There are many (hyperbolic and misleading) examples:

> At Amazon we are committed to sustainability and invest in it because it is beneficial for everyone: for the planet, for the business, for our customers and for our communities.[39]

> We [Apple] create products and services to enrich the lives of our customers. And we strive to do so in a way that sustains the planet and the resources that we all depend on.[40]

> The ultimate success of our business [Uber] will depend on our ability to transition our platform to clean energy in collaboration with drivers, industry innovators and governments. It's the right thing to do for our customers, our cities, our shareholders and the planet we all share.[41]

Despite proclaiming their devotion to green and sustainable practices, neither Amazon, Apple, nor Uber (among others) have reduced their emissions. In fact, they have multiplied them. A brief analysis of their own sustainability reports—in which they claim to have reduced their energy consumption, fossil fuel emissions, and raw material use—highlights the fallacy of green and digital capitalism. For example, Amazon has gone from emitting 5.57 million metric tons of carbon dioxide equivalent (CO_2-eq) in 2019, to emitting 12.11 million in 2021.[42] Even in a context of crisis and market saturation, Apple continues to sell hundreds of millions of devices every year. It would be unfair to ignore the fact that technological advances are improving efficiency, for example the use of less water and the increasing use of recycled materials in Apple's latest iPhone and Samsung's Galaxy models. However, improving efficiency per unit is not the same as reducing emissions or resource consumption.

Not surprisingly, the World Bank forecasts that by 2050, the production of minerals such as graphite and cobalt could increase by 500 percent.[43] As denounced by the Obersavatorio de las Multinacionales de América Latina (Latin American Multinationals Observatory) research center, the new wave of extractive mining megaprojects looming on the world's peripheries is due precisely to the "green and digital transition."[44] Every new phone sold, every new data center, every new transatlantic submarine cable has an ecological footprint that depends on the extractive-industrial processes described above. Capitalism cannot grow without causing harm and, as Marxist criminologist David Whyte comments, ecocide is inscribed in the corporate DNA.[45] Above all, these companies are committed to the ultimate truth of capitalism: perpetual growth. For example, Amazon has stated that it will increase its fleet of electric vehicles from "more than 1,000" to 100,000 units by 2030. The green and digital revolution is instigating a new global extractive turn, in which geopolitical struggles for control of critical resources demanded by cutting-edge technologies will become increasingly common—more mines, more extractive machinery, more trailers, more containers, more cargo ships, more processing centers, and more battery production industries, all in the name of the ecological transition.

Consider, for example, the Australian-funded San José Lithium mine, planned just two kilometers from the historic city of Cáceres (Spain).

With great cynicism, the corporate narrative has portrayed the mine as a sustainable and ecological project with little or no negative environmental and social impact.[46] Only the resistance of the population has temporarily halted illegal drilling and logging and the construction of new roads. In the same vein, the multiplication of wind farms in Ireland—intended to meet (not reduce) the skyrocketing energy demand of data centers—is falsely presented as an example of sustainability and decarbonization.[47] Campaigns encouraging compulsive consumption of the latest trendy gadgets are greenwashed and presented as ethical consumption; the multiplication of infrastructures to support the green and digital transition is presented as the virtuous path to economic growth while combating climate change. The problem is that this is simply not true.

The Religion of Infinite Growth

Walter Benjamin referred to capitalism as a religious phenomenon. Lacking theology and dogma, it nevertheless claims a permanent and total worship "sans rêve et sans merci [without sleep and without pity]."[48] One of the main gestures of the capitalist liturgy is the public expression of faith in an impossible and therefore almost eschatological fact: perpetual growth in a world with limits. As André Gorz explains, the idea of economic growth hides, behind its appearance of neutrality and simplicity, complex and ideologically charged theories.[49] First, the concept presupposes the economy as an autonomous sphere dissociated from other spheres, such as the social, political, or ecological. Growth in production, sales, consumption, and capital flows is seen as the metric of success, regardless of its environmental or human impact. Second, within this ideological model, everything must be quantifiable, with a value that can be accounted for in monetary terms. The economy is reduced to what the logic of the market can apprehend, thus turning money into the absolute equivalent of a five-hundred-year-old tree, a carbon credit, or a smartphone. It is a spell of black magic, by virtue of which everything becomes a commodity or is expelled from the economic realm.

The size of market operations is equated with progress and development. The larger the trade—whether it is organic apples produced by a worker-owned organic farm or the activity of a large open-pit coal mine—the better. This is true for micropolitical actors as well as for states. An example of this ideological framework is the ubiquitous metric of gross domestic product (GDP), the monetary value of goods and services produced within given borders. GDP has become the global measure of success, regularly used to compare the "development" and "progress" of some countries with respect to others. For example, the United States had a GDP of $23.32 trillion in 2021, China of $17.73 trillion, and Brazil of $1.609 trillion. GDP is usually divided by the total number of people residing in a given territory to obtain GDP per capita, a supposed indicator of the well-being of the population in that territory. But there is a catch. Only what falls within the sphere of the market is considered "wealth";[50] what does not (yet) have a monetary equivalent, such as domestic work, care, wildlife, clean air or personal fulfillment outside of wage relations, does not count. The consequence of this mentality is obvious. What is quantifiable and monetizable must grow and multiply, for that is the currency of the common good, even if it devours everything else (computers and microchips versus mountains, batteries for Teslas versus Andean salt flats). The prevailing growth narrative, which often camouflages the commodification and exploitation of nature and humanity, receives unanimous praise from leading advocates of the ecological and digital transition, regardless of their political or economic affiliation.

> Economic growth is going up, stronger than experts expected, at 2.9 percent we are growing. [Joe Biden][51]

> Let us rediscover the spirit of Maastricht: stability and growth can only go hand in hand. [Ursula von der Leyen][52]

> Our Consumer revenue grew dramatically in 2020. In 2020, Amazon's North America and International Consumer revenue grew 39% . . . [this trend] extended into 2021 with revenue growth of 43% year-over-year in Q1 2021. These are staggering numbers. We've realized the equivalent of three years of forecasted growth in about 15 months. [Amazon][53]

> We continue to advance important growth opportunities and projects to bring to market critical materials for the transition to a low-carbon economy: copper, lithium and iron ore, among others. [Rio Tinto][54]

There is no doubt that the idea of economic growth espoused by, among others, Joe Biden, Ursula von der Leyen, Jeff Bezos, and Jakob Stausholm has translated into increased production and purchasing power for both large organizations and individuals, especially those in the Global North. But it is also true that this capitalist myth legitimizes the destruction and consumption of the Earth's material base. Should we safeguard a mountain and its resources or exploit these resources and turn them into minerals and disposable items (with an eco-label, of course)? The Rio Tinto mining company is clear about its allegiance, and it is not to the planet. How else can it be explained that, in a context of global climate and environmental crisis, the corporation's annual reports to shareholders celebrate exponential growth in extraction and applaud its gigantic profits? Specifically, there was an increase in the extraction of all the minerals they preach, which has led to a 116 percent increase in profits compared to 2020, from $9.8 billion to $21.01 billion in 2021, and a 70 percent increase in dividends to shareholders over the same period.[55] How much growth is enough to satisfy the greed of shareholders? How much can the planet endure? Little.

According to (conservative) data from the United Nations International Resource Panel, global raw material use increased from 26.7 billion tons per year in 1970 to 75.6 billion tons in 2010. In 2020, this amount exceeded 100 billion tons for the first time and is expected to reach 190 billion tons by 2060.[56] To put things in perspective, the total global biomass—referring to the weight of each plant, animal, fungus, protist and monera—is estimated to be about 550 billion tons. In 2020, we consumed the equivalent of one-fifth of that amount, and our hunger continues to grow. Every year, we consume more biomass, fossil fuels, metals, and minerals, not only in global terms, which go beyond the increase in population (which was 3.7 billion in 1970 and 7.8 billion in 2021), but also in terms of per capita consumption.[57]

According to the Global Footprint Network, human voracity for resources has for several years now exceeded the Earth's capacity to replenish them (biocapacity). To determine the biocapacity of a given area, the Global Footprint Network has devised a calculation that measures the amount of biologically productive land and sea area available to meet people's demands and absorb their waste, taking into account current management practices and technology. The result is expressed in global hectares (gh). Our world has an estimated total of 12.2 billion gh, or 1.6 gh per capita. However, the global distribution of consumption is not homogeneous. The per capita ecological footprint of people living in the United States is 8.1 gh, or five times this amount. The footprint of a person living in Spain is 4.3 gh, in Russia 5.3 gh, in Argentina 3.3 gh, in Bolivia 1.9 gh, and in Iran 3.3 gh. Moreover, it is worth noting that the material footprint per capita from high-income countries reveals an average of 26.3 metric tons per capita per year.[58] In stark contrast, those residing in low- and middle-income countries exhibit significantly lower figures, 2 and 4.7 metric tons per capita per year, respectively. This glaring disparity underscores the profound discrepancies in resource consumption across socioeconomic and geopolitical contexts. Today, humanity consumes the equivalent of what 1.75 Earths could produce;[59] we would need five Earths to consume sustainably at the average rate of the US population.

Let's look at one of the most notable victims of human-caused devastation: forests. Humanity has gobbled up about 35 percent of all forests, half of this in the last 120 years alone. Between 1700 and 1850, 1.9 million hectares of forest were cut down each year, rising to 3 million hectares per year in the period between 1850 and 1920. Between 1920 and 1980, this number increased to 12 million. The rate of deforestation has slowed in some regions of the Global North (or rather, has been delocalized), but even taking this into account, 101 million hectares of forest cover were lost between 2000 and 2020. Approximately 29 percent of the world's land area is covered by forests (in contrast to 56 percent in prehistoric times);[60] this figure includes fragmented, secondary, and tree-crop forests destined for logging. Only 25 percent of forests can be considered intact, and a meager 10 percent are primary forests, that is, the forested areas where the highest levels of biodiversity on Earth are concentrated. Primary forests are virtually nonexistent in places like Europe, where only a few remnants of what were once massive primeval forests survive as beleaguered relics threatened by logging. Most of the remaining primeval forests in the Global South are threatened by transnational forestry, food, and mining corporations. As a painful example, between 2019 and 2022, 45,585 squatre kilometers of Amazon rainforest were sacrificed to the gluttonous god of capital.[61] According to a recent study, about 10 percent of rainforest deforestation is caused by industrial mining, such as the Carajás mine in Brazil, which is the largest iron ore mine in the world.

Such destruction undoubtedly affects animal life. To paraphrase Francisco de Goya, the dream of global capitalism produces monsters. For years, the NGO World Wildlife Fund (WWF) has been producing the Living Planet Index, which attempts to estimate global changes in wild animal populations. According to its latest report, the world has lost almost 68 percent of its wildlife between 1970 and 2022. These figures rise to 94 percent in the case of Latin America. For example, the population of marine sharks and rays has declined by 70 percent since 1970, and two-thirds of these species are now at serious risk.[62] In fact, nearly one million plant and animal species are threatened with extinction. Species such as the four swamps mojarra, the Zumarraga and the Antipodean albatross may soon join the Caspian tiger, the California brown bear or the Pyrenean ibex on the long list of the sixth extinction. While this information is certainly not new, having been debated and ignored for decades, it seems to have escaped the attention of the powers that be. Neither mass extinction, nor the felling of forests, nor the blowing up of mountains, nor the capture of rivers, nor the destruction of the seas are enough to mobilize high politics.

It has been the climate emergency, and its more than palpable social, political, and economic repercussions, that has escalated climate and environmental discussions to corporate and state forums, for example, discussions on climate change in the United Nations system, where, despite frequent disagreements, the relationship between human action and the global rise in temperatures is widely recognized.[63] Whether from an ecocentric point of view (the Earth, the natural environment is in danger) or an anthropocentric one (our country, region, or civilization is threatened), a new common sense recognizes the existence of a deep and widespread global problem. See, for example, the speech of the president of the European Commission, Ursula von der Leyen, at the Earth Day 2021 summit of world leaders: "The Paris Agreement is humanity's life insurance. At COP26 in Glasgow, we must show that we have all understood this and that we are ready for further climate action. Because we are getting dangerously close to 1.5 degrees of global warming. Science tells us that it is not too late, but we must hurry up."[64]

Hurry up, of course, but how and to where? What are the recipes proposed by global capitalism to tackle the climate crisis? Despite the enormous political, geographic, and social distance that separates them, international organizations, interregional governments, large and small countries, and companies embrace the same slogan: our societies must make the transition to a green and digital economy. The International Monetary Fund—the great champion of global neoliberalism responsible for, among other things, Greek austerity—now advocates the need to move "towards a green, inclusive and digital future."[65] The European Union speaks of the "dual green and digital transition."[66] Telefónica and Cisco refer to the green and digital transitions as a single phenomenon. It is, as a *Forbes* headline states, "how shell is using Web3 and blockchain for sustainability and energy transition."[67] Meanwhile, Rio Tinto, perhaps one of the most environmentally damaging corporations worldwide, proudly proclaims to have launched the "first sustainability label for aluminum using blockchain technology."[68] In short, the institutions responsible for global environmental destruction are proposing a turbocharged, digital, green version of capitalism as a solution to their ecocidal crimes.

Market Techno-Solutionism

For years, the European Union has been cultivating an image of modernity and progress. It has presented itself as the last bastion of democracy in a fractured world, a territory where citizens can exercise their political rights, where markets have limits, and where the fight against climate change (without renouncing technological progress) is a priority. This is what makes its techno-solutionist narrative a valuable case study. Two of the European Union's flagship strategies are the twin transitions: "A Europe fit for the digital age" and the "European Green Pact." In their own words, "The green transition aims to achieve sustainability and combat climate change and environmental degradation," while "in the digital transition, the European Union aims to harness digital technologies for sustainability and prosperity, and to empower citizens and businesses."[69] This is no mere rhetorical exercise; it is an ambitious economic recovery plan that aims to finance the green and digital reindustrialization of Europe, using €1,211 billion from multiannual funds and €806 billion from new generation funds. Eighty percent of these funds are to be specifically earmarked for the dual green and digital transition, which is seen as Europe's moonshot in the fight against climate change, or the "cornerstone for delivering a sustainable, fair and competitive future."[70] The message, tone, and discourse developed by the European Union are clear. Climate change can only be tackled with technological means, industrial modernization, digitalization, and replacement of fossil energy sources with renewables. It is paradoxical to see how the techno-industrial society and its socio-technical elements are presented as the saviors of the planet given the adverse effects they have inflicted on it.

This narrative merges the idea of sustainability and economic growth into a single message, as if the two dimensions were dependent and indistinguishable. For the European Union, it is perfectly possible to maintain the current economic model and level of consumerism while at the same time restoring the environment and the biosphere. The European Commission even offers its own version of a fully automated, digital, ecological, and capitalist future:

> As you wake up in 2050, you might begin your day by looking out of the window, with your augmented reality device showing you real-time pollution data. You then have breakfast that you bought because you were convinced by its environmental score, which was clearly visible at the time of the purchase, thanks to digital data. The food itself is produced by farmers in a resource-efficient manner, because they know exactly which crop to plant when—they have access to Big Data, thanks to open-source platforms gathering public environmental information, weather forecasts, or data through on-farm sensors. Before you turn on your washing machine, you check the electricity price at the moment. To incentivise consumption at periods when renewable-produced energy is abundant, the prices vary, and gamification is making the hunt for good timeslots a fun experience. You not only consume but also produce electricity, thanks to the solar panels installed on your roof, which is connected to a meshed micro-grid.[71]

Despite the eco-futuristic references and the explicit homage to the American New Deal, the object of these funds is neither public services nor citizenship articulated in sustainable eco-cooperatives. The real beneficiary of the billions in public funds will be the private sector in charge of carrying out the green and digital reindustrialization of Europe. But who is the private sector that would selflessly carry out the double transition? In October 2022, one of Europe's many lobby groups published a report outlining how Europe should operationalize its green and digital transition. The issue would be irrelevant were it not for the fact that behind the *Action Plan for a Digitally Enabled Green Transition* lurk some of the world's most powerful corporations.[72] A bit of history is in order.

On April 17, 1983, seventeen of Europe's most powerful capitalists created the European Round Table of Industrialists (or ERT), a business lobby with an explicit mandate to influence European Union decisions. Its members include the mining companies Rio Tinto and ArcelorMittal; the arms manufacturer Leonardo; the chemical company Badische Anilin- und Sodafabrik (BASF SE); the cement company Alexandria-Portland Titan; the energy companies British Petroleum, E.ON, and Iberdrola; as well as large communications companies such as Vodafone, Orange, and Telefónica (among many others). Many of these companies have been implicated in serious corporate, environmental, and human rights crimes. Rio Tinto supported the bloody 1936 fascist coup in Spain and subsequently supplied essential minerals to Nazi Germany during World War II. Its extractive activities remained during the twentieth century and beginning of the twenty-first century the germ of environmental devastation, repression, assassinations, and civil wars such as the Papua New Guinea-Bougainville war.[73] Today, the corporation's operations and prospections are resisted by peoples around the world, one of the most notable examples of popular resistance being against its lithium extraction megaprojects in Serbia. British Petroleum was responsible in 2010 for what may be one of the worst environmental disasters in North America and (so far) the largest oil spill at sea: the Deepwater Horizon oil spill. ArcelorMittal and Titan have been found responsible for the deaths of dozens of people as a result of emissions from their factories; the former, owned by the UK's richest man, is involved in a giant iron ore mine in the Qikiqtaaluk Region, a traditional Inuit territory in Nunavut (Canada). The inhabitants have been protesting for years against the mine and projects to expand it.[74]

These criminal enterprises have privileged access to high circles of European power. In January 2023 alone, ERT met four times with members of the European Commission Cabinet responsible for digital transition, internal markets, and financial affairs. This amounts to 126 meetings during the 2014–23 period.[75] The fact that this megacorporate lobby has met dozens of times with senior European officials at critical moments in policymaking is no accident. Nor is it a coincidence that among the major beneficiaries of European funds earmarked for decarbonization and the ecological transition are players from the old fossil fuel industry such as Iberdrola, Repsol, and Petronor, which—far from abandoning their old revenue streams—have seen their profits multiply in recent years.

The scene would not be complete without the Big Four: Deloitte, PricewaterhouseCoope, KPMG, and Ernst & Young. These historic advocates of public-private initiatives have been advising governments for years in strategic areas such as healthcare, communications, security, and energy policies. An illustrative case is highlighted by Ekaitz Cancela and Stuart Medina, who reveal that the Spanish state has allocated a staggering €378 million in ten years to the four large corporations for advisory services related to the public communications agency and the railway network.[76]

The European Commission has also frequently called on their services, with nearly €462 million spent on advisory services in the 2016–19 period alone.[77] Aware of the immense possibilities opened up by the new European funding channels, consultancy firms have quickly adapted to the green capitalism narrative, replicating a clear and unambiguous message: green and digital capitalism will allow us to stay on the path of economic growth without giving up the fight against climate change. For example, the report *The Twin Transition: A New Digital and Sustainability Framework for the Public Sector,"*commissioned from Ernst & Young by Microsoft, insistently repeats a mantra: digital technologies are the key to achieving total sustainability. Amid dozens of images of virgin forests, mountains, and rivers, the report describes how, thanks to digital capitalism, the public sector will be able to move towards a "sustainable society and economy; sustainable municipalities, cities and regional governments; monitoring, modeling and management of the Earth's natural system; and sustainable government."[78] Without a doubt, the consulting firms come out on top. On the one hand, the Big Four are hired by governments at all levels to collaborate in the design of sustainable development and digital transition strategies; on the other, the consultancies show technicians the way to generously funded European public tenders. This is, of course, a textbook example of conflict of interest—but who cares?

The European Union, like other territories of the Global North, boasts of having reduced its emissions and pollution levels while increasing its forested areas as a clear symptom of the adequacy of its solutionist policies and the efficacy of technological change. However, this eco-facade does not hide the fact that the outsourcing of "externalities" such as manufacturing, extractivism, and waste to the Global South explains the "greening" of the EU. Beyond the propaganda, the truth is that there is no digital and green recovery without an industrial society organized around fossil fuels. A case in point is coal-fired factories being driven off European soil only to be relocated to China. Regions like Europe simply cannot consume at the current unbridled level without preying on external resources, be they materials or labor. It seems that the only way to save humanity (i.e., the capitalist economy in the form of green industry) is by exploiting the reserves of rare minerals—the new ecological and digital transition is not halting the disastrous course of events but accelerating it.

Civilization or Barbarism: European-Style Green Imperialism

Despite the divergent sentiments that the green and digital transition arouses among different political sensibilities—fervent green deal advocates, right-wing denialists, technocratic greys, collapsists—all recognize the foreseeable and monumental escalation in demand for mineral resources. To mention one example, according to the European Commission, European demand for lithium, graphite, and cobalt could increase by as much as forty-five times, fourteen times, and fifteen times, respectively, by 2050.[79] This demand will result in a titanic effort that will require the opening of hundreds of extractive and processing megaprojects, not to mention the vast network of infrastructure linked to them. Political documents such as the "European Green Deal" have framed this destructive process within the discourse of sacrifice, responsibility, and the need to take forceful measures to achieve "climate neutrality by 2050," a utopian path that "aims to transform the EU into a fair and prosperous society, with a modern, resource-efficient and competitive economy where there are no net emissions of greenhouse gases in 2050 and where economic growth is decoupled from resource use."[80]

But the situation is certainly more complex and less sweetened. The contemporary geopolitical reality marked by the undeniable rise of powers such as China and India and the growing international conflicts provoked to a large extent by imperialism, such as in Russia and Ukraine, Syria, Yemen, and Palestine, have highlighted the fragility of powers such as the United States or the European Union, which are dependent on the resources, technologies, and supply chains needed to keep the digital era afloat. The relative scarcity of critical resources in rich countries is recognized by the central countries and clearly pointed out by the lobbies of digital capitalism as a fundamental problem that threatens not decarbonization and climate neutrality (something secondary) but the constant and assured flow of resources that guarantee both competitiveness and security. To cite well-known facts, the EU imports 98 percent of its rare earths from China, 68 percent of its cobalt from the Congo, and 78 percent of its lithium from Chile.[81] And these are just some of the many examples that show how the critical infrastructures of the "powerful" depend on the permanent plundering of the oppressed.

One of the most striking features of green extractivism is the ease with which it has recovered old imperial discourses, which have historically legitimized environmental destruction, the undermining of Indigenous sovereignty, the strengthening of the white-colonial supremacist regime, and colonization as a necessary price to pay for civilization.[82] The words of the UN Economic Affairs Officer for Europe, Olga Algayerova, at the United Nations Climate Change Conference meeting in 2022 are not far from the Spanish imperial argument used for centuries: salvation in exchange for your resources. Algayerova said, "The world is mired in a deep energy crisis and needs an urgent energy transition. However, this transition cannot take place without massive amounts of critical raw materials (CRMs) needed to deploy the low-carbon technologies required for climate change mitigation and adaptation."[83]

European Commission president Ursula von der Leyen expressed similar sentiments in her 2022 State of the Union address: "If we do not have secure and sustainable access to the necessary raw materials, our ambition to become the first climate-neutral continent is in jeopardy."

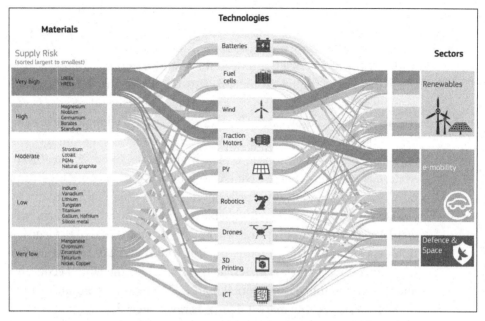

Figure 5.2: "Semi-quantitative representation of flows of raw materials and their current supply risks to the nine selected technologies and three sectors." Bobba et al. 2020, 10.

Von der Leyen at least had the decency to admit that the issue of resource control was not only for the sake of the environment, but also for the sake of the military industry and, more importantly, the economy: "Whether we are talking about chips for virtual reality or cells for solar panels, the twin transitions will be fueled by raw materials. Lithium and rare earths are already replacing gas and oil at the heart of our economy. By 2030, our demand for these rare earth metals will increase fivefold. And this is a good sign, because it shows that our European Green Pact is making rapid progress."[84]

This institutional discourse is accompanied by a corporate narrative of an unabashedly extractive nature. In particular, the prominent Digital Europe lobby, which counts Google, Amazon, Microsoft, and Apple among its members, fervently supports the commission's aggressive green extractivist measures, advocating more capitalism and exploitation while urging the reduction of what it describes as excessive red tape and slowness in the granting of mining permits. In fact, the industry has gone a step further by demanding the elimination of regulatory barriers for new mining megaprojects, which it demanded be considered strategic projects of public interest[85]—something that, as I will show, they finally obtained.

Motivated by these arguments, the European Commission decided to renew the regulatory framework for strategic mining to align it with the principles of digital capitalism. This was a legislative process that, as Friends of the Earth Europe revealed in an intelligent report,[86] was strongly influenced by extractive and metal industry lobbyists. The official announcement of the new EU legislation leaves no room for ambiguity about its intentions and priority interests, titled "Critical Raw Materials Act: Securing the New Gas & Oil at the Heart of Our Economy."[87] This legislation appropriate to the new green imperialism draws a double extractivist line, a double plundering gesture: outward and inward.[88] For what purpose? The first line of the first recital of the legislation exposes with crisp brutality the imperialist and extractivist nature of the text: "Access to raw materials is essential for the Union economy and the functioning of the internal market."[89] No excuses, no subterfuge, no half-measures. The market demands raw materials.

In its first gesture, the new imperial-extractivist law seeks to safeguard the access of the rich countries of the EU to the resources of the Global South through trade and investment agreements. This is made clear in countless texts which, like the communication accompanying the legislation, even propose the creation of a "Critical Commodities Club": "Access to a secure, affordable, and sustainable supply of CRMs is a shared concern among many partners and a key theme in many intergovernmental fora (such as G7, G20, the International Energy Agency and the International Renewable Energy Agency). . . . The EU should complement and build on these various actions to establish a Critical Raw Materials Club bringing together consuming and resource-rich countries to promote the secure and sustainable supply of CRMs."[90] This is not a statement of intent but a political one. In parallel to the drafting of this legislation, the EU launched a diplomatic offensive in search of resources. Perhaps the most representative is that of lithium, which I will also discuss in the following chapter. Most of the world's reserves are located in Andean territories with a majority Indigenous and peasant population. In this context, at the end of 2023, the European Union signed a vast trade agreement with Chile in which lithium was the real protagonist. Despite the fact that the European Union has signed most of the international commitments to human rights, and more specifically to Indigenous populations, there is no mention of the fact that this mineral is mostly located in the ancestral territories of Native populations. It is striking to see how, despite the prolific production of regulations, communications, and scientific reports, the EU ignores in its production that not only lithium, but also that most of the strategic mineral reserves, are located in Indigenous and peasant territories. Once again, international law, mobilized under the narrative of civilization and redemptive progress, gives way to a new extractivist era at the service of the global north. This new law erases Indigenous jurisdictions, contributes to the abolition of ancestral territories, and proves to be the sword and shield of both the extractivist corporations and the white-creole elites of the Global South.

In its second gesture, the new wave of legislation opens the door to new extractive projects in the peripheries of the Global North. The new European legislation sets as a target that at least 10 percent of the new projects should be satisfied with extractions coming from European soil. At first sight, it might seem that this is news with a certain fairness; after all, the European predatory vocation is usually dedicated to the peripheral territories "beyond enemy lines" where there seems to be no right, where everything is allowed. But this superficial analysis does not hold water. First of all, if capitalist logic teaches us anything, it is that it only trusts in infinite and exponential growth. That is to say that this hypothetical 10 percent of resources extracted from European soil will not replace the resources extracted in Africa, Asia, and America, but will be added to them to manufacture more goods. On the other hand, when Europe speaks of mining, it is not referring to the rich regions, or to the foothills of the great metropolises that will end up consuming the products. The areas to be sacrificed in Europe, such as the French Massif Central, the Balkan valleys, or the mountains of Galicia, Portugal, and Extremadura, are rural areas, territories with a strong peasant tradition, which are among the most impoverished and marginalized areas in Europe.[91] They are already subjected to enormous extractivist mining, wind, forestry, and water pressures. They are ultimately a periphery subjected to internal colonialism.

The new regulatory framework not only legitimizes but also obliges countries to draw up a program for the exploration of critical raw materials. It also obliges the authorities to prepare the legal infrastructure for planning, zoning, and proper management of extractive projects conceived as critical and strategic. In other words, the new legislation determines extractivism as a priority for European development and welfare. Following the demands of capitalist lobbies, the new legislation streamlines and smoothes mining administrative processes, allowing, for example, environmental self-certification schemes and relaxing the due diligence of mining companies. The new scheme accelerates the granting of strategic permits and places mining exploitation above other issues of general interest. This prioritization of extractive land use accompanies criminal legislation aimed at protecting critical projects and infrastructure—a process of dispossession that anticipates the criminalization of social protest.

Finally, the legislation is tailor-made for financialized extractivist capitalism, for which a decisive role is foreseen and which will also be subsidized with public funds such as the Next Generation funds that are already being channeled to mining and automobile companies, among others. This legislative and industrial route is precisely the one being followed by regional and national governments, such as those of Extremadura and Portugal, to ensure the exploitation of their lithium reserves. All this is despite the frontal opposition of communities that have long been forgotten, all in the name of the future of humanity.

Far from offering alternatives to the consumerist model that brings us closer to climate catastrophe, the green and digital transition demands the acceleration of extractive, colonial and industrial logics. In doing so, it resonates with the infamous imperial narrative: your resources in exchange for our civilization and salvation. Beyond the rhetoric and fictional environmentalist narratives, in this chapter I have shown how there is a firm alignment between the state and the interests of capital to ensure the flow of raw materials from the Global South to the Global North in order to safeguard the economy and the function of the market. In other words, in order for rich regions to remain competitive, grow economically, and subsidiarily decarbonize their economies, it will be necessary to continue to clear forests, mine valleys and mountains, and dissolve tons of materials into hyperpolluting chemicals that will be left as a legacy to Indigenous and rural communities. In the end, as always, there remains a dialectical question based in a brutal relationship of contradiction—which will we as a society choose: capital or life?

6

Killing the Salar de Atacama

What once seemed like a radical gesture—claiming and recognizing the rights of nature—has now become part of the twenty-first century's political and legal common sense. Seeds cultivated from long-standing struggles have been planted. A growing consensus asserts that the Earth should be afforded the highest levels of protection, and that it has an intrinsic spiritual and metaphysical value that reaches beyond its instrumental utility as a material resource. This sentiment is echoed by an increasing number of voices taking a strong stance against the aggression (some would say war) of states and corporations against nature. Immediately, a myriad of questions arise: How can life that exceeds an anthropocentric conceptualization be protected through legal and political instruments? How can the rights of nature be applied and defended? There are no easy answers or simple solutions to this problem of global dimensions. Nevertheless, for a growing number of institutions, social movements, political forces, and communities, the first step is to criminalize the most blatant and calamitous offenses against the environment. This chapter situates the environmentally harmful crimes of digital capitalism within these debates.

There are many ways to start a story, none of which are innocent or neutral. If I were to adhere to the conventional path, I would commence by discussing the environmental and conservation movements of the nineteenth century followed by the ecological shift that unfolded in the mid-twentieth century. I would delve into the nonanthropocentric resituation that thinkers like Christopher D. Stone eloquently expounded uon in 1972. I would then examine how Rafael Lemkin, the originator of the term *genocide*, laid the groundwork for explicating what would later be known as ecocide.[1] I would discuss how the context of the Vietnam War, alongside US social movements and theory, led to Arthur W. Galston's work successfully introducing the concept of ecocide into the international sphere, particularly in defining some of the atrocities committed by the US Army.[2]

This canonical genealogy would continue to highlight the new wave of ecological movements that emerged in the early decades of the twenty-first century in the Global North (such as Extinction Rebellion[3]) which was fueled by increasing evidence of climate change. I would speak to the bravery of a group of activists who managed to bring the issue of ecocide from the margins of expert opinion to the forefront of contemporary criminal environmental policy debates. This narrative unfolds effortlessly, carrying the familiar and self-satisfied undertones found in American blockbusters focused on justice. It conveys a reassuring message, reminding us that even in the face of our personal and collective errors, significant strides towards structural transformation are attainable through the implementation of some legal and political reforms. After all, Stop Ecocide—the organization created by the late activist Polly Higgins—had managed, from its humble regional origins in the 2010s, to offer a legal definition of ecocide that has been discussed by states and international institutions.[4] This story is nice, but it is far from a complete picture.

Some of the most innovative, promising, and inspiring environmental legal instruments that lay the groundwork for criminalizing ecocide did not originate from the naturalist sensitivities of the Global North. As I will show, these achievements are embedded in historical processes of resistance against intersecting forms of domination, including colonialism, racism, anthropocentrism, patriarchy, and capitalism. Law is not, as many would have it, a mere product of economic structure, nor are the social rights we enjoy today the product of a vanilla dialectic. Every civil, political, economic, and environmental right has been wrested from power at a terrible cost, and it is misleading to claim that the expansion of national and international environmental law, and even the criminalization of the most serious offenses against nature, will by itself change the destructive mechanics in which we are immersed. There is no change without a mass movement—without a profound transformation of the mental and material structures upon which our societies are built. Social conquests die if organized peoples are not there to protect them. Still, I find the notion of ecocide rich, provocative, and politically useful. It provides a grammar from which to name some of the most intolerable forms of state and corporate environmental harms. In the following pages, I will critically engage with the concept of ecocide to grasp the highly destructive, yet routinized operations of digital capitalism.

Green, Southern, and Decolonial Criminology

One of the most interesting debates in criminology today has to do with questioning the capacity and even the legitimacy of a discipline that was born hand in hand with racist, colonial and policing interests. To cite just one of many examples, in 2021, the Munanjahli and South Sea Islander activist and writer Chelsea Watego, in one of the keynote speeches at the University of Melbourne's Criminology Department (the oldest criminology department in the country), proposed "making the case for abolishing criminology."[5] She questioned the discipline for its complicity in three hundred years of colonialism, genocide, and ecocide. Two years later, the criminologist Maria Giannocopoulos reiterated these arguments, detailing the complicity of colonial legality in the regime of dispossession imposed by white supremacy on the Australian continent. Maori criminologist Juan Tauri showed how criminology contributed to the mythologisation of white civilized subjectivity, serving the interests of colonial criminal justice. Tauri, following in the footsteps of decolonial criminology pioneer Biko Agozino and Maori Moana Jackson (himself a pioneer of Indigenous criminology without calling it that), has shown how criminology has collaborated closely with the criminal justice system in the development of social control sciences aimed at Indigenous oppression and the erasure of non-Western justice systems. These voices challenging the disciplinary and academic perpetuation of the colonial capitalist structure have been joined by a wave of sharp new critiques from Indigenous and racialized writers such as Amanda Porter, Natalie Ironfield and Jasmine Barzani. What purpose, then, can be served by a science so intimately linked to the instruments of power that have led to the genocide and ecocide of dozens of peoples and territories? As the decolonial school of criminology has pointed out, the discipline, despite its obvious limitations (which must necessarily lead to its timely abolition), also offers adequate tools to uncover and question the complex mechanisms that underlie the assembly of capitalist-colonial structures of oppression.

Proof of this is the emergence of approaches such as decolonial criminology, green criminology, or criminology of the South, which make it possible to analyze and situate in its complexity the socio-environmental devastation perpetrated by states and corporations today. For example, authors such as José Atiles, David Rodríguez Goyes, Juan Tauri, and Valeria Vegh Weis have shown how the global logic of capitalist accumulation, which benefits a handful of "powerful" people, depends on the destruction of Indigenous peoples for its operability and affirmation.[6] This logic of annihilation is articulated in a twofold movement.[7] First, it pursues the symbolic-material and cultural undermining of Indigenous peoples and nature, attacking, for example, the legitimacy of Indigenous sovereignty, the rationality of the relationship with nature, and the civilisation of the relationship of these peoples with the territories. It applies relationships to these relationships the logic of civilisation and development that we have already seen. Second, the capitalist-colonial logic of accumulation calls for the material destruction of Indigenous peoples and the material base with which they are intertwined. This is what some define as the genocide-ecocide nexus.[8]

The ruling classes have long been aware of the symbolic power of legal form in the production of reality. This is why they privilege some legal forms while pursuing the destruction of others. We need only look at the history of that legal aberration, that fetishization of capital, known as the joint-stock company (see chapter 1), which is an artificial legal entity, albeit fully functional, with a wide range of nationally and internationally recognized rights that are privileged over the rights of nature and of Indigenous and peasant peoples.

The resistance of Indigenous, peasant, and rural peoples shows that the struggle for the rights of nature and territory goes beyond the narrow confines of Western legal utilitarianism. As Anne Spice points out, it is a complex phenomenon, a tangle of relationships and life processes at the intersection of the legal, the natural, and the infrastructural.[9] Activist and thinker Leanne Betasamosake Simpson beautifully and poetically describes this intricate web as a confederacy that transcends anthropocentric boundaries, linking human and nonhuman nations in relationships of coresponsibility. She sees "nations as networks of complex, layered, multidimensional, intimate relationships with human and non-human beings. Our societies work very well when those relationships are balanced. Our legal system isn't a series of "laws" based on authoritarian power with punishments for when the laws are broken. It is an embedded and interwoven spiritual, emotional, and social system of intelligence that fosters independence, community, and self-determination in individuals."[10]

Defending mother earth, defending the rights of nature, is a way of defending life in its biological dimension, but it is also, in the same movement, a gesture of resistance that shields the confederation of peoples. This double movement, which is both defense and attack, questions by its very existence the hegemonic symbolic space of the state and capital where the limits of what is politically desirable or possible are defined. When the First Nations of the "Red Mountain" mobilize to prevent the construction of a lithium mine in Thacker Pass, they are not only defending the land, their ties, and their rights as a people today, but the very capacity to inhabit, produce, and defend horizons that are not capitalist, colonial, anthropocentric, or patriarchal, which have been threatened for centuries.[11]

The Struggle for the Rights of Nature

Forest is one of my favorite words; it signals a world in just a few letters. I have familial, political, romantic, academic, and historical memories attached to forests. I try to spend as much time as I can under the treetops, sometimes above the upper tree line, reaching the high peaks, but usually not far from there. Each forest, regardless of its familiarity, possesses its own unique allure—its magic, its heart. Two forests stand out in my memory for vastly different reasons. Lizardoia, a forest of beech and old oak trees nestled in Euskal Herria,[12] stirred up a sense of familiarity; the combination of its rain, trees, sounds, smells, leaves, and rocks made me feel at home. Yet, Te Urewera (Aotearoa [New Zealand]) was an entirely different experience. There was something indescribable about that place—something ancient, colossal, and powerful. I could not help but be drawn to its profound presence, which evoked a deep sense of respect within me. I remember the majesty of the tōtara[13] trees: solid, mysterious, proud, and resilient. Among the branches of these conifers—survivors from the ancient supercontinent of Gondwana—live kereru,[14] kokako,[15] and kākā.[16] Their songs bounce off the trunks of rimu,[17] tanekaha,[18] and matai.[19] When the sunlight manages to penetrate the tangled foliage, it takes on a greenish, golden hue. The dark, calm, and serene allows for the growth of a wide variety of ferns, from the elegant kiokio[20] to the huge whekī-ponga[21] that resemble trees. Moss covers the ground, the roots, the fallen trunks, and the branches. Moisture saturates every inch of the place. The rivers flow joyfully, swiftly, and yet somehow peacefully and anciently at the same time. In them, the whio—a species of endemic duck with a dark-bluish color—quack with their distinctive sound. There are fewer than three thousand left, and they all call this forest home. The presence of Te Urewera stands out on the North Island of New Zealand. It does so because the New Zealand "green branding" that presents the islands as a natural paradise fails to hide the calamitous devastation caused under the colonial rule of the Pākehā, the name the Māori gave to the white settlers. In less than fifty years of imperialism, colonialism, and capitalism, these settlers destroyed nearly 50 percent of the forests that existed before the first Māori waka[22] arrived on the islands.[23] This is double what the Māori population had caused in a thousand years of learning and coexisting with the forests.

The fact that Te Urewera has survived to this day is not due to the benevolence of the colonial Pākehā regime or modern environmental awareness. Te Urewera has survived thanks to the iwi[24] that has inhabited it for hundreds of years, and precisely because they have defended it and turned it into a place of struggle against colonialism. Ngāi Tūhoe have been one of the most oppressed peoples under the colonial regime. The Tūhoe were not part of the founding (myth) of the New Zealand nation, which occurred at the signing of the Treaty of Waitangi in 1840 between the British Crown and some rangatira[25] of Te Ika-a-Māui (North Island of Aotearoa). The Tūhoe remained committed to rangatiratanga,[26] Te Reo Māori,[27] and kaitiakitanga[28] over Te Urewera.[29] The Ngāi Tūhoe and the forest form a way of life, where the human, the social, and the natural intertwine through rituals, hunting, shared kai,[30] and herbal healing. The Tūhoe do not exploit or manage Te Urewera: they are the forest, and the forest is them.

This political-ethical framework and the set of situated knowledges that emanate from territorialized practices is what theorist of the Dene Nation Glen Coulthard and scholar of the Michi Saagiig Nishnaabeg Nation Leanne Betasamosake Simpson refer to as grounded normativity.[31] This is a form of self-governance and knowledge that is at the same time a way of being and of inhabiting the world. The Tūhoe resisted in the nineteenth century. They challenged imperialistic wars, legalized land plunder, and subsequent famines. They fought state sponsored epistemicide and genocide. In the twentieth century, the Tūhoe defied the persecution of their leaders, the Great Depression, industrial logging, the expansion of intensive farming, and aggressive development. Well into the twenty-first century, they had to face their vilification as a people, the neoliberal wave, and the criminalization of protest. In 2007, the New Zealand state launched one of the largest police operations in its recent history. Hundreds of militarized forces assaulted rural Tūhoe communities in search of an alleged Indigenous terrorist insurgency, during which seventeen people were arrested.[32] This operation also extended to Christchurch, Wellington, Taupo, and Tauranga, where the police raided social centers and houses of allegedly dangerous environmental activists who stood in solidarity with the defense of Te Urewera and the Tūhoe cause. In 2012, the Crown courts dismissed most of the charges, and in 2014, the police chief publicly apologized for the excess force of the police operatives.[33]

In that same year, the New Zealand government and the Tūhoe Iwi reached an agreement that, as activists recognize, did not end the conflict but opened a path to follow. Symbolically, the state recognized the historical violations of rights, land theft, and organized violence by the state against people, land, and culture. The state agreed to provide financial, economic, and commercial reparations. It also committed to the restitution of Mana Motuhake or self-governance and sovereignty of the Tūhoe.[34] In the wake of this agreement, the Te Urewera Act emerged, which recognized the legal personality of the forest; it did so not based on protectionist or conservation arguments that had historically been used to displace Indigenous populations from their lands, but by recognizing that

> Te Urewera is a place of spiritual value, with its own mana[35] and mauri.[36]
> Te Urewera has an identity in and of itself, inspiring people to commit to its care.
> For Tūhoe, Te Urewera is Te Manawa o te Ika a Māui; it is the heart of the great fish of Maui, its name being derived from Murakareke, the son of the ancestor Tūhoe.

For Tūhoe, Te Urewera is their ewe whenua, their place of origin and return, their homeland.

Te Urewera expresses and gives meaning to Tūhoe culture, language, customs, and identity. There Tūhoe hold mana by ahikāroa;[37] they are tangata whenua[38] and kaitiaki[39] of Te Urewera.[40]

In 2017, the Te Urewera Act was followed by the recognition of the Whanganui River and Mount Taranaki,[41] which resulted from processes of struggle, defense, and resistance. Rather than being isolated incidents, these processes are situated in a lineage of Indigenous struggles for the rights of nature. A few years earlier and thousands of kilometers away, Ecuador and Bolivia had introduced in their constitutional texts the protection of the rights of Pachamama (Mother Earth) and Sumak Kawsay (Good Living), both concepts emanating from the grounded normativity of the Aymara and Quechua peoples. These concessions were not benevolent gestures from the ruling class, but the result of revolutionary processes. Both South American countries have a long history of colonial and postcolonial extractivist policies—policies that have enriched white and creole minorities, leaving a legacy of environmental destruction, Afro-slavery, and Indigenous servitude.[42] This structure of economic inequality and political power has defined politics in the region from the colonial era to the collapse of neoliberal regimes in the early 2000s.[43] It was at this moment that Indigenous, Afro, and mestizo sectors—along with urban and rural social movements—managed to balance their contradictions to dispute and attempt to transform state institutions that had caused the plunder, devastation, and destruction of their communities and territories for centuries. As noted by theorists of the so-called new Latin American constitutionalism, these emancipatory processes aimed to radically transform Latin American legal reality.[44] One of the most outstanding elements of these processes was the recognition of pluralism and interculturality.

Historically, Latin America has been a region characterized by the political, economic, and cultural hegemony of a white and creole minority over a rich and diverse majority. In the legal realm, this has been evidenced by the ongoing tension between different sources of legitimacy, sovereignty, and legal authority. For example, in the community of Xochipala (Mexico), located in the so-called gold belt, Indigenous peoples, the Mexican state, narco, and mining corporate legalities coexist in constant conflict. This multiscalar or legal pluralist setting has historically been subjected to the interests of the dominant white and mestizo class, which has not hesitated to use lawfare to plunder the resources of Indigenous and peasant peoples.[45] For authors such as Carlos Wolkmer, Luis Tapia, and many others, the project of the New Latin American constitutionalism, which was driven—at least to some degree—by historically oppressed subjects, proposed a different, transformative, and emancipatory horizon of possibilities.[46] New Latin American constitutionalism offered a participatory possibility capable of recognizing the historical, material, and immaterial demands of peoples, such as their sovereignty as peoples and the spiritual rights of forests, moors, valleys, and mountains.[47] It offered a legal horizon that placed the communal-popular, rather than the individual, at the center. This is, in the words of Carlos Wolkmer,[48] a legal horizon inspired by emancipatory rationality.

In the history of oppressed peoples, the legal and constitutional realm has been an extraordinarily important site of struggle. Let us think, for example, of the first constitution of the emancipatory American continent. The Haitian constitution had already abolished slavery in 1801, relegating it to the dustbin of history. This was at the very moment that slaveholding colonial states, led by white minorities in Brazil and the United States, were multiplying the theft and trafficking of human beings.[49] The Mexican constitution of 1917 was one of the first in the world to recognize broad social rights, and the Soviet constitution which promulgated a year later was a milestone, both within and outside the socialist field, inspiring a whole new generation of collective rights.[50] There are countless other legal examples both at the national and international levels[51]—and there are many more examples outside the realm of Western legalism, such as in the Indigenous tools of self-governance, sovereignty, and law, which have been used to produce territories of resistance as well as to protect nature and the peoples who inhabit it.

However, as countless critical voices have highlighted, the constitutionalization of some of the rights of nature and the mere recognition of legal personality to forests, rivers, and mountains does not guarantee their protection.[52] This is why I believe that a grammar such as that of the ecocide is necessary to point out the aggressions of the powerful. I certainly do not believe that regulation will succeed in stopping the destructive logics of capital. Nor do I believe, with the punitivist confidence, that the more or less real threat of punishment will deter the destroyers. I do believe, however, that the criminalization of ecocide is a starting point from which to operate, from which to name some despicable facts, and from which, perhaps, to launch insufficient, but still necessary, strategic litigation—a legal guerrilla operation that undoubtedly has to be accompanied by real changes, more vast and profound, and at the same time granular and extensive.[53]

Towards a Definition of Ecocide

Environmental destruction is a phenomenon that knows no borders or territorial limits. Let us consider the Great Pacific Garbage Patch, a massive current of microplastics that spans over 1.6 million square kilometers between the coasts of Japan and the United States. Artificially designed to persist, plastics will sail, fracture, and float for centuries without decomposing. Not only is this a nuisance, but microplastics pose a risk to countless marine species, affecting their growth and reproductive capacity, among other things.[54] Microplastics also threaten humans: although the full range of consequences are still unknown, the United Nations Environment Programme has highlighted that microplastics have entered the food chain. We do not have to go further beyond the seas to find another human-made catastrophe. According to data from the Food and Agriculture Organization, 36 percent of fish stocks are being overexploited, a percentage that rises to over 60 percent for Mediterranean and South Pacific regions. As noted by blue criminology, both legal and illegal fishing industries are driving numerous species to the brink of extinction, whether for commercial interests or due to destructive extraction techniques. Iconic species that once populated the seas in huge numbers—such as the Atlantic salmon, the Atlantic cod, or the albacore tuna—are now threatened.[55] Oceans are not only home to millions of species, they are also the source of the oxygen we enjoy on Earth and are responsible for capturing a significant portion of the CO_2 in the atmosphere.[56] At the same time, the meat industry in the United States has been found responsible for the Gulf of Mexico dead zone, a marine region of 6,334 square miles where low oxygen levels make it incompatible with life.[57] In other words, mining, carbon emissions, deforestation, and the plunder and destruction of rivers and seas causes devastating damage while simultaneously threatening the regenerative capacity of the Earth.

The international nature of these crimes is expressed not only through their brutal global consequences, but also through the actors who cause them. Each of the "singular" environmental disasters of the past twenty years has been produced by transnational corporations. Take for instance the cyanide spill caused by an Australian-Romanian joint venture that contaminated the rivers shared by Romania and Hungary,[58] the Texaco oil spills in Ecuador,[59] or the deforestation of the Amazon rainforest in Brazil to make way for intensive livestock farming.[60] Certainly, the problem is not limited to a handful of worst-case scenarios. If there is one thing that academic and scientific forums agree on, it is that the routine actions of public and private actors—that is, day-to-day activities—are the main source of global emissions and devastation. We know that environmental destruction is a problem that affects the Earth, and that the actors mostly responsible for this destruction operate in various jurisdictions. So, what international tools exist to address these attacks?

In an extraordinarily judicialized world covered by complex legal frameworks, one would expect legislation that addresses this threat to our shared ecosystems. However, as recognized institutionally and academically, there is a significant legal gap in international law.[61] Global society currently lacks international criminal-legal instruments aimed at protecting the Earth. It is not that attempts have not been made. In 1995, members of the United Nations called for the creation of an international convention that would address the greatest threats to the environment. In 2001, the UN Environment Programme (UNEP) established an intergovernmental group of experts to develop guidelines that would bring clarity to different multilateral environmental agreements in order to promote the consistent and accurate implementation of these agreements. Yet the most that has been achieved is the obtaining of soft law formulas that are merely declaratory and have little international legal value.[62] In fact, there is not even an international consensus on how to define international environmental crime. The closest definitions are those offered by UNEP-Interpol and Europol, respectively:

Although the definition of "environmental crime" is not universally agreed, it is most commonly understood as a collective term to describe illegal activities harming the environment and aimed at benefiting individuals or groups or companies from the exploitation of, damage to, trade or theft of natural resources, including, but not limited to serious crimes and transnational organized crime.[63]

Environmental crime covers the gamut of activities that breach environmental legislation and cause significant harm or risk to the environment, human health, or both.[64]

If we descend to the regional level, we find at least one legal instrument aimed at dealing with international environmental crime. The Directive 2008/99/EC of the European Union was intended to be the EU's silver bullet against the fourth largest source of crime, with an annual growth rate of 5–7 percent and one of the highest rates of impunity.[65] However, according to EU reports, this initiative has been a criminal policy failure (2020). According to the EU's data, the number of reported cases of environmental crimes to the European Union Agency for Criminal Justice was extremely low, with only fifty-seven cases registered between January 1, 2014, and December 31, 2018. We are talking about a region with twenty-seven member states and 447 million inhabitants; these statistics represent less than 1 percent of the total cases handled by Eurojust during that period. Moreover, the cases were only reported by sixteen of the twenty-seven countries, with the majority of these cases committed in the Netherlands, France, and Germany. In a thorough report, the EU's criminal agency analyzed several socio-legal problems that have hindered the proper implementation of the directive.[66]

Firstly, the EU acknowledged that there is a preference for the application of administrative law rather than criminal prosecution, resulting in few cases reaching the prosecutor's office. This is further aggravated by weaknesses in cooperation among environmental inspectors, police, and prosecutors, which results in a lack of data and mechanisms for data exchange. These challenges have led to fewer cases being processed, with further problems arising in evidence and perpetrator identification. As recognized by the EU, the lack of specialized prosecutors and the low priority given to environmental crimes due to a lack of resources (time, money, expertise) has also contributed to the limited number of cases initiated, with even fewer brought to courts. Furthermore, even when cases *are* brought to court, there are issues with evidence and a lack of knowledge in environmental criminal law, resulting in dismissals on grounds of expediency. As with prosecutors, there is also insufficient specialization among judges in cases of environmental crimes, which leads to most convictions resulting in lenient sanctions. This is further complicated by the fact that environmental crimes are often considered less serious than traditional crimes. Based on this, the EU has proposed a new, stricter, and more comprehensive directive to protect the environment through criminal tools.[67]

It was in the context of this impunity that Stop Ecocide's proposal to criminalize the most onerous state and corporate environmental harms emerged. Until her passing in 2019, Polly Higgins, Stop Ecocide's founder and leader, was one of the most visible advocates for including the crime of ecocide within the jurisdiction of the International Criminal Court. The activist defended her position before numerous forums, including the United Nations, but it is within the European legal system that her legacy has gained the greatest traction. Parallel to the European debates calling for new green criminal legislation, Stop Ecocide launched an international panel of experts in 2020 tasked with creating a legal definition of ecocide. The committee offered the following definition in 2021: "For the purpose of this Statute, 'ecocide' means unlawful or wanton acts committed with knowledge that there is a substantial likelihood of severe and either widespread or long-term damage to the environment being caused by those acts."[68]

This definition has received widespread media, political, and institutional reception, particularly in the EU. In April 2023, it was announced that the European Parliament proposed including ecocide in the new European directive aimed at protecting the environment through criminal law in the following terms: "Member States shall ensure that any conduct causing severe and either widespread or long-term or irreversible damage shall be treated as an offense of particular gravity and sanctioned as such in accordance with the legal systems of the Member States."[69] While trilateral discussions between the Parliament, the Council, and the European Commission are ongoing, it is expected that ecocide will be included in the final version of the legislation. Furthermore, it is expected that the directive will be significantly more severe than the current law for both natural and legal persons. Though specific figures for sanctions are yet to be defined, the discussed range includes fines of between 3 percent and 25 percent of the total average worldwide turnover of the legal entity, as well as prison sentences for natural persons. Standing on the theoretical ground outlined earlier, as well as the legal definitions I have just described, I will analyze a thoroughly documented case of lithium mining operations in the Salar de Atacama, Chile. My aim is to demonstrate that the extractive operations demanded by global digital capitalism are responsible for the ecocide of a unique ecosystem and the subsequent social annihilation of the people who have lived there for thousands of years.

Killing the "Desert"

Atacama is an uncertain term, its exact origin unknown. It could refer to the word used by the Quechua people to name a black duck, the yequ, which is common in the Atacama wetlands, or the terms for "cold" or "village" in the Cunza language spoken by the Lickanantay people. Whether it means duck, cold, or village, Atacama has become part of the global language of geopolitics and technology through countless articles, reports, and high-level discourse. Atacama is situated at one end of what has been branded the "lithium triangle"—a region split between Argentina, Bolivia, and Chile that produces around 30 percent of the world's resources and could hold around half of the global lithium reserves.[70] Lithium is on the list of critical materials—that is, essential for national security and economic development—of two of the world's three most important economic blocs: the European Union and the United States.[71] This relatively abundant mineral is essential for the batteries that power almost all of our electronic devices and, more importantly, it is indispensable for the so-called decarbonization of transportation. Goldman Sachs estimates that by 2040, sales of electric cars will rise to 73 million annually.[72] To put this in perspective, according to the International Energy Agency's conservative estimates, each electric car uses an average of 8.9 kg of lithium.[73] Some people think that technological change will allow for smaller batteries, but the fact is that even though we are capable of producing smaller batteries, we are also producing larger and more powerful cars, which in turn require more minerals. This is not to mention the enormous demand that the decarbonization of the global industrial machinery could have. The war over lithium control is in the cards.

Lithium may be a driving force behind a new green and digital economy. If we look at the model(s) that govern lithium exploitation and pay attention to those who benefit and those who are harmed by its mining processes, it is difficult to differentiate it from the old colonial model of accumulation by dispossession.[74] Certainly, the political processes opened up by Indigenous resistance as well as progressive governments have allowed for a timid redistribution of exploited wealth,[75] for example, by forcing corporations to establish negotiation processes with Indigenous communities (as in Chile) or establishing extractive public entities (as in Bolivia). But even under the progressive or liberal neoextractivism regime, states and corporations continue to monopolize benefits while peoples and nature carry the burden of destruction.[76] Drawing from the insights of decolonial thinkers such as Arturo Escobar and Hector Alimonda, it becomes evident that the coloniality of nature surpasses ideological and territorial boundaries.[77] Within this context, the three dominant modern states—Bolivia, Argentina, and Chile—perpetuate a systemic pattern that bears a striking resemblance to their colonial predecessors, viewing the region primarily as a source of exploitable resources. This exploitative mindset is exemplified by the Chuquicamata copper mine, a vast and deep open-pit mining operation renowned as one of the largest in the world. The mine's tunnels, projected to reach an astonishing depth of approximately eight hundred meters during peak production,[78] epitomize the historical legacy of resource extraction in the area. Throughout history, the wealth of this territory has been extracted by various entities, including the Incas, Spaniards, English, Americans, and Chileans. Notably, even the Guggenheim family, known for its philanthropic endeavors in the realm of art, amassed a fortune by capitalizing on the devastation and exploitation of this region.[79]

Despite Bolivia's vast potential reserves and Argentina's persistent and destructive efforts to exploit lithium, Chile leads the region's production. Chile alone contributes 20 percent of global production, most of which is concentrated in a small area of about three thousand square kilometers known as the Salar de Atacama basin.[80] Two corporations monopolize lithium extraction in the area. First, there is Sociedad Química y Minera (SQM) a former public corporation privatized during Augusto Pinochet's fascist regime in Chile and today's second-largest global producer.[81] Its production in 2022 is estimated to have reached 180,000 metric tons equivalent of lithium, a figure it expects to increase (along with its profits) to 210,000 tons by 2025. Second, there is Albemarle, a US conglomerate that co-owns the lithium mine in Greenbushes, Australia. In 2022, Albemarle predicted that it would extract 140,000 metric tons of lithium from Chile.[82] Regardless of which capitalist investment fund benefits from this dispossession, the destination of the mineral is the same: China. In 2021, 60,782,000 out of the 134,060,000 tons of minerals extracted in Chile ended up in China.[83] In fact, the two corporations that monopolize lithium exploitation in Chile maintain an intense relationship with the megacorporation Tianqi Lithium (present alongside Albemarle in Greenbushes Australia), which controls about half of the world's lithium production.[84]

SQM and Albermarle refer to the Atacama Salt Flat as a "desert" on their corporation's websites, a sacrificial zone with no use other than mineral extraction.[85] The idea of the desert as an abandoned, depopulated, uncivilized, and lifeless place has populated the colonial imaginary since its origins. Not far from the region, the genocidal Campaña del Desierto (Desert Campaign) (1878–90) took place, in which the Argentine State mobilized its army against Mapuche and Tehuelche populations, decimating them almost to extinction.[86] The notion of desert as a vacated place was used widely under the ignominious colonial doctrine of *terra nullius*, which justified the extermination of First Nations people and brutal environmental devastation in what is now Australia, Canada, and the United States.[87] The concept of the desert, far from its geographic neutrality, delineates a topography of what is conquerable, dominable, what can be taken and ultimately destroyed. The Atacama Salt Flat is undoubtedly an arid territory where precipitation is a relatively meager 27.8 millimeters annually. But Atacama is not a desert (in the colonial sense of the word); it is a space of life. In this fragile ecosystem, thousands of endemic species coexist. The shy rivers and streams that descend from the Andes, as well as the hidden masses of groundwater that emerge in the spine of America, nourish a complex water network whose outcroppings allow for the migration and mobility of vicunas, pumas, condors, and flamingos.[88] That this is a territory full of life and deserving of protection has been made clear by the numerous protected natural spaces that dot the territory, such as the National Flamingo Reserve in the salt flat basin.

The salt flat is also home to human societies. The Lickanantay people have been inhabiting these territories for at least five thousand years. Their words and practices, their stories, the archaeological record, and the widespread rock art attest to this.[89] The Lickanantay's way of life—based on agriculture and animal farming—revolves around water, which is scarce yet present. Their populations are distributed around outcroppings and springs, their rituals and worldview are intertwined with water and its channels,[90] producing what is known in anthropology as a hydrosocial territory.[91]

The Salar de Atacama is far from a pristine territory. This hydrosocial interweaving of nature, water, and human life has been in tension due to the mineral wealth of the territory. The Lickanantay people have survived the mining ambitions of the Incas and brutal Spanish colonialism; they maintained their identity and ways of life based on agriculture and livestock during the Chilean industrial revolution built on copper mining; they navigated the authoritarian neoliberalism of Pinochet that unleashed the privatization of water; and they weathered the liberal multiculturalism of the 1990s by articulating new forms of organization, such as the Atacameño People's Council (CPA).[92] It was the resistance of the CPA and other peoples that forced the Chilean state to negotiate the recognition and return of some of their ancestral territories in the 1990s and 2000s. Unfortunately, this recognition did not extend to the most precious entity: water.[93]

Despite this history of political repression, it is the green and digital transition that is pushing the material basis of the Lickanantay's way of life to its limit. Lithium in Atacama is not found in solid mineral form. Instead, it flows in the form of brine through the underground veins. This brine is pumped and channeled into thousands of pools where—through different and lengthy processes (up to fourteen months) of natural drying and chemical purification—the mineral and other by-products such as alite, sylvanite, sodium chloride, and carnallite are obtained.[94] SQM is authorized to extract between 1,500 and 1,700 liters per second (l/s) of brine, while Albemarle is authorized to extract 442 l/s. In addition to this, both companies are authorized 263.5 l/s of fresh water.[95] This is an extraction of about sixty-three billion liters of water per year, as much as the entire population of the Antofagasta region would need, far exceeding the recharge capacity of the aquifers. To these legal figures, we must add the hundreds of thousands of liters of water illegally obtained by mining companies like Albemarle, which is currently under investigation for overexploiting aquifers.

Lithium Ecocide

In 2018, a report commissioned by the Production Development Corporation (CORFO)—the Chilean agency in charge of ensuring "economic development"—confirmed what Lickanantay populations had long been denouncing: the balance of the lagoons was being threatened. According to the data in the report, the recharge capacity of the Salar basin (calculated at 6,975 l/s), was clearly exceeded by extractive flows, which ranged from 8,442 l/s to 8,842 l/s.[96] In other words, the data provided by public institutions (and therefore presumably conservative) indicates that legal and illegal practices linked to lithium mining are causing profound and irreversible damage to the water reserves on which the natural life and hydrosocial fabric of the Salar depend. To this data, we must add the countless studies carried out by social organizations and universities that point to the discussion of river and reserve flows, the drying up of wells, and the subsequent effects on plant, animal, and microbial life. We must further add those studies that are looking at the impact on Indigenous and peasant social structures. Wenjuan Liu and Datu B. Agusdinata, from the University of Arizona, have demonstrated the degradation of the basin, especially in the areas closest to the mining. Specifically, they have found accelerated deforestation and loss of cultivated areas, a decrease in soil moisture, and an increase in surface temperatures.[97]

It is particularly ironic, although not surprising, to note how companies responsible for the legal and illegal exploitation of water in the region present themselves as corporations committed to the "preservation, and restoration of the area's ecosystems."[98] Highlighting this contradiction, Gonzalo Gajardo and Stella Redón ask, "How will Chile combine biodiversity agreements, treaties on wetlands, and endangered bird conservation under the soaring lithium demand to support electromobility?"[99] Their study indicates that the increase in salinity in the lagoons caused by mining overexploitation is altering the fragile balance of the ecosystem and reducing the number of small organisms and invertebrates on which birds depend. The reduction in available food, as well as suitable nesting space, is putting the James's flamingo and the Andean flamingo at risk, whose reproductive success has already dropped with a population decline of 10–12 percent.[100] The region is also the last refuge for threatened species such as the colocolo cat, the guanaco, and the critically endangered short-tailed chinchilla.[101]

I have pointed out, the lagoons are not just a natural space, but are part of the broader hydrosocial fabric inhabited by the Lickanantay people. Both their material base, centered on agriculture (alfalfa, corn, and fruit trees) and the breeding of camelids, as well as their grounded normativity are articulated around water.[102] Lickanantay are distributed around the freshwater outcrops, establishing affinity links with them. The salt flats are not a resource; they are an entity that allows life in the region. This is evident in the collective cleaning of the canals organized by the Lickanantay ayllus (Andean traditional political units), which merge religious rituals, community configuration strategies, and the pragmatic distribution of collective work necessary for survival.[103] The mining impact on water resources, both those suitable for irrigation and human consumption and those that are saline, goes beyond the mere reduction of a resource. Extractive capitalism undermines a fundamental pillar of the practices, rituals, and production spaces that shape the social life of an ancestral community. The destruction of the Salar is thus an attack to the Lickanantay people *as* people—an expression of what many critical scholars have denominated as the genocide-ecocide nexus.[104]

On a legal level, several organizations have been denouncing the fact that both the Chilean state and corporations repeatedly fail to comply with the International Labour Organization Convention 169, which includes the right to prior consultation—free and informed—on decisions that affect Indigenous peoples.[105] The United Nations Declaration on the Rights of Indigenous Peoples (UNDRIP) signed by the Chilean state is also consistently violated by state and corporate practices.

Among the rights violated by the extractive operations are the right to self-determination, the right to determine development priorities, and the rights to land and territories,[106] as well as Articles 25 and 29 of the UNDRIP, which state the following:

> Article 25: Indigenous peoples have the right to maintain and strengthen their distinctive spiritual relationship with their traditionally owned or otherwise occupied and used lands, territories, waters and coastal seas and other resources and to uphold their responsibilities to future generations in this regard.[107]

> Article 29: Indigenous peoples have the right to the conservation and protection of the environment and the productive capacity of their lands or territories and resources.[108]

Unlawful or Wanton Acts

Some authors have pointed out that the Chilean extractive regime is, thanks to its regulatory framework, less burdensome and more benevolent than the neoliberal brutality of neighboring contexts, such as Argentina. But this is misleading. First, Law No. 20.600 established Environmental Courts in Chile, which are responsible for ensuring compliance with environmental permits. This legal resource has been widely used by Chilean indigenous peoples to denounce the serious environmental destruction of their territories which has resulted, at best, in punitive sentences that barely make a dent in the profit of corporations. A clear example of this is the investigation that Albemarle is facing for the overexploitation of water resources in the Salar, for which it could be fined only US$500,000.[109] The Chilean state has also overseen the renovation of mining concessions for SQM and Albemarle which has led to a well-known agreement between Albemarle and the Lickanantay people, whereby the former must transfer 3.5 percent of their sales to the community. Beyond the questionable success of this operation, the fact remains that the Chilean state continues to disregard Indigenous self-government over their territories. The populations of the Atacama Salt Flat Basin have been and continue to face an asymmetric war of dispossession of their ancestral territories.

Despite the meager transfers made by mining company Albemarle and the sanctions imposed by the state, the destructive mining of the Salar de Atacama, which is the material foundation for the existence of the Lickanantay people, continues. Corporations are pushing for its acceleration by demanding the expansion of projects and new extraction sites, such as that proposed in the nearby Salar de Maricunga. In other words, mining corporations defend not only the legality but also the legitimacy of serious violations of both the rights of nature and the Lickantany people. I would like to recall for a moment the definition of ecocide given by the independent group of experts convened by the Stop Ecocide Foundation: "Unlawful or wanton acts committed with knowledge that there is a substantial likelihood of severe and either widespread or long-term damage to the environment being caused by those acts." Given this definition and the established facts, it is clear that lithium mining in the Salar de Atacama meets the requirements to be considered a case of ecocide—an ecocide necessary for the routine operations of digital capitalism.

While identifying these types of environmental crimes is an essential step towards potential solutions, it does not address the most critical questions: What actions should we take, and how can we achieve them? Are polities such as the EU moving in the right direction? Personally, I am skeptical. While the European Parliament is advocating for the criminalization of serious environmental offenses, it is also approving a new extractivist regulation, the Critical Raw Materials Act,[110] which will accelerate environmental devastation under consumerist, colonial, developmental, and capitalist logics. These contradictions give rise to more pressing questions: What value can environmental (or any other) legislation have when the legal architecture reinforces a regime that is antagonistic to life? To what extent can a punitive approach that relies on state structures inscribed in the text of capitalist and colonial oppression address the harms it has inflicted for centuries? These are the questions I will address in the conclusion of this book.

Conclusion

Platformed Criminals

Whenever I participate in a conference or talk about digital capitalism, I always start with the same questions. Who in the room has a mobile phone with internet access? Who has an operating system based on Android or iOS? It doesn't matter if the talk takes place in a small town or a big city, whether it's in a country from the center or the periphery. It doesn't matter the what the audience's socioeconomic level, cultural background, or religious and political beliefs are; the answer is always the same. An overwhelming majority of attendees have access to a device with internet access. I have never found anyone using a different operating system than Apple iOS or Android. The attentive reader may rightly point out the profound bias of this on-site observation, devoid of any scientific value, or flag how numerous studies demonstrate the enormous structural, economic, and social challenges that historically marginalized groups, such as the First Nations of Australia or elderly people in rural Spain, face when accessing digital technologies. I won't deny proven facts such as the digital divide, however, despite these exclusions and marginalizations (reflecting historical processes of discrimination), the global trend is moving towards a forcibly capitalist and digital world.

Let's look at some data. At least 58 percent of the population in Senegal has internet access, a number that rises to 72 percent in South Africa, Egypt, and Gabon and 88 percent in Morocco. Virtually 100 percent of the population in countries like Saudi Arabia, Qatar, and Bahrain have internet access. In Asia, 73 percent of the population in China is connected, while 74 percent and 98 percent are connected in Vietnam and South Korea, respectively. In Bolivia, 66 percent have access, while it's 81 percent in Brazil, 87 percent in Argentina, and 90 percent in Chile. Internet access in the European Union averages at around 84 percent.

Interestingly the majority of these connections are made through mobile devices.[1] This list could go on to cover nearly all the countries in the world, and despite the significant differences between territories and regions, the number of connected people continues to increase.

Never before in human history has a single technology achieved such extensive territorial and population reach, approaching quasi-universality in just three decades. According to archaeological studies, technologies such as bows and arrows took at least ten thousand years to spread from Africa to Europe.[2] Agriculture, originating in American, Mesopotamian, and Asian centers, diffused through a slow process that required hundreds of generations.[3] Despite the progressive acceleration of knowledge and technology exchanges (and impositions), technologies such as steam power, internal combustion engines, telegraphy, and telephony took close to a century to reach critical mass. Facebook took only 8 years to reach its first billion users.[4] According to the latest available data (2022), around 2.96 billion people use an application from the Meta conglomerate daily, and that number increases to 3.74 billion people when considering monthly users.[5] In other words, nearly half of the global population frequently interacts with their applications. In fact, even today, infrastructure considered common in early twentieth-century London, such as electrification or gas, is still absent in numerous places where, surprisingly, one can enjoy an endless stream of videos and posts.

There is no doubt. A ghost haunts the world—the ghost of technologization, digitalization, and hypercommunication. It is a swift spirit that rides on the back of the most dystopian turbo-capitalism. In just fifteen years, from the 2007 launch of the iPhone to the present day, global societies have unquestioningly and naturally embraced the need to carry a device at all times and in all places that allows (and compels) individuals to be available, locatable, and communicable. It is a machine through which users document every second of their existence, access leisure and pleasure, and even essential services of daily life such as welfare or banking.

The digital has become routinized, made invisible by merging its infrastructure with economic and social materiality. But it is precisely these socially accepted facts, these circumstances taken for granted as part of new common sense, that should draw the attention of inquisitive minds. One of the things that surprised me the most upon arriving in Aotearoa-New Zealand was the incredible ubiquity and penetration of social networks (especially the Facebook ecosystem). Any cultural, musical, or artistic event, no matter how alternative, seemed to necessarily have its counterpart on Facebook events. Similarly, political organizations, from mainstream parties to university groups and abolitionist organizations, used the platform not only as a showcase but also as a vital tool for coordination and organization. This dependency extended to interpersonal relationships, where Messenger (on Instagram and Facebook) had replaced calls and SMS in a considerable segment of the population. In other words, anyone in Aotearoa-New Zealand with even a minimal cultural, political, or purely social interest had to engage with the techno-social system owned by the Meta conglomerate.

For the vast majority, this doesn't represent a significant change or a major problem compared to other technological developments. The growing social dependence on the digital world would be a small price to pay for the development of a set of technologies that, according to their defenders, facilitate communications, promote commerce, and enable the transition to a greener, fairer, and more solidary economy. Why should we protest against the progressive cybernetization of bodies and societies? After all, there are few voices complaining about agriculture, roads, electrification, or communication networks, for example. Following this logic, the rapid advancement of digital capitalism is not a change of course but rather the next stage of development in the techno-scientific civilizational project brilliantly described by Lewis Mumford.[6] This widespread acceptance of the developmental ideology, shared by both the left and the right, in the Global North and South, does not mean that the growing social dependence on digital technologies is free from problems, frictions, and questions. The explanatory reason has more to do with the actors who organize and manage this dependence. It means public and private institutions wield power, often irresponsibly, over these powerful socio-technical systems.

Let me go back to the example of Meta. Brazil, a country with 212 million inhabitants, has 134 million active Facebook users, 120 million on WhatsApp, and 72 million on Instagram (of which 64 percent are between eighteen and thirty-four years old). These individuals spend around three and a half hours per day interacting on social networks.

For half of the population of this country, Facebook, WhatsApp, and Instagram are the main source of news and the gateway to reality.[7] For some, this is seen as another manifestation of US imperialism,[8] while others view it as a new form of domination labelled data colonialism.[9] In any case, the potential political power that such capillarity and media reach could irresponsibly use is undeniable. And that's precisely what happened during the 2018 Brazilian electoral process.[10] For months, the Silicon Valley company ignored the calls to attention denouncing the homophobic and racist misinformation campaign of the far right, which would ultimately lead to the election of Jair Bolsonaro as the president of Brazil.[11] What came after that was a neoconservative backlash that accelerated the destruction of the Amazon, eroded Indigenous and Black people's rights, and threatened the lives of LGBTQIA+ communities after state-promoted hate speech campaigns.

As noted in numerous sources, the Facebook-Meta conglomerate has become a platform for amplifying xenophobic, authoritarian, sexist, and racist discourses.[12] For instance, legal scholar Neriah Yue has highlighted the potential legal responsibility of Facebook-Meta in relation to the Rohingya genocide in Myanmar.[13] This historically marginalized Muslim minority endured a brutal state-led campaign of violence in 2016, with Meta's social platforms playing a pivotal role in coordinating these crimes against humanity. Despite initially evading responsibility, the company eventually admitted in 2018 that they had not done enough to halt the violence in the country.[14] However, this apology fell short for the victims of the conflict, who have pursued legal action against the company in British and American jurisdictions, citing the corporate complicity in these crimes.[15] It is worth noting that these concerns are not exclusive to Meta.

The major corporations involved in the rapid rise and expansion of digital capitalism have been repeatedly held accountable for causing significant social harm. Alphabet, the parent company of Google and Android, for example, has faced legal repercussions in numerous jurisdictions for systematically violating privacy regulations, engaging in anticompetitive practices, infringing intellectual property rights, and failing to adequately protect children's data.[16] Executives at Uber infamously claimed, "We're just fucking illegal," shedding light on the routine disregard for legal and ethical boundaries.[17] This tsunami of (so-called) scandals, which I consider to be more accurately labeled as corporate criminal practices, have spurred public scrutiny and led to calls from political institutions to address the role of large technology corporations in society. Consequently, a multitude of legal and legislative initiatives have been undertaken worldwide in an attempt to curb the power of big tech, as I will elaborate on later. However, this chapter seeks to delve deeper by exploring not only the question of "What is happening?" but also "What can we do?"

My intention in this conclusion is to initiate a dialogue that invites the necessary critique of digital capitalism and its crimes, while also fostering the production of alternatives to it. I will embark on this process from an unconventional standpoint—critical criminology. The starting hypothesis (which I have previously defended at length with James Oleson[18]) is that the most influential technology companies (such as Amazon, Apple, Alphabet, Microsoft, and Meta, among others) have achieved their dominant positions through systematic and deliberate violations of the rules of the game, particularly privacy and competition regulations. This has resulted in profound and extensive social harms. In other words, I argue that the relationship between digital corporations and crime is not incidental but rather structural, and, more specifically, their business model relies on crime, whether or not it is codified in bourgeois legal systems. To illustrate this, I will focus on some of the social harms caused by a well-known company, Meta, and examine some of the (insufficient) measures taken by Global Northern states to mitigate these harms. Specifically, I will critically analyze the European Digital Services Act, which may be the boldest attempt to govern digital platforms. The final sections of the chapter present some thought-provoking ideas, pointing towards hypothetical transformative scenarios in which critical criminology could potentially be of assistance.

From Data Harms to Data Crimes

On September 14, 2021, the *Wall Street Journal* published the first of what would soon be known as the Facebook Files.[19] An anonymous source, later revealed to be Frances Haugen, a former employee of Facebook-Meta, leaked highly compromising documentation about the tech giant. According to the leaked documents, Meta had conducted dozens of experiments to understand the psychological impact on its users. The results of these experiments were clear: the mechanisms behind products like Instagram, designed to capture and maintain user attention, had toxic consequences for their mental health. In her testimonies before the US Senate Subcommittee on Consumer Protection, Product Safety, and Data Security (2021) and the European Parliament (2021),[20] Haugen stated that Instagram was designed to maximize the company's profits, with little consideration for the consequences it could have on users and society at large.

For instance, the algorithms developed by Meta have inadvertently contributed to the perpetuation of mainstream Western beauty standards, which prioritize material wealth, whiteness, youthfulness, and extreme thinness in women. One noteworthy example is the application's provision of a wide range of filters explicitly designed to modify users' images, primarily targeting young women. These filters offer options to lighten users' skin, slim down their bodies, or make them appear younger. Beneath the seemingly innocuous facade of these applications lies a technology that influences adolescents to view their bodies as imperfect and incompatible with the heteropatriarchal, white, and Western beauty ideals. Such tools impose a subtle form of violence on users' bodies and minds, creating significant disparities between their actual physicality and the algorithmic standards of beauty crafted by the digital capitalism industry. The case brought to light by Haugen serves as an exemplification of what critical literature refers to as data harms or "the adverse effects caused by uses of data that may impair, injure, or set back a person, entity or society's interests."[21]

The extent of the psychosocial damage caused by these systems is still not fully understood, but preliminary data confirms that young people who spend more time on social media platforms like Instagram are more prone to experiencing mental health issues, self-harm, and even suicide. According to psychologist Jonathan Haidt from New York University, there is a clear causal relationship between the alarming rise in depression cases and the erosion of young people's mental health and the use of social media.[22] Unlike other high-profile scandals involving tech giants, authorities from various jurisdictions called Haugen to testify. For Haugen, the issue was clear. "Don't trust them," she said in her testimonies, adding that they are putting "astronomical profits before people."[23]

On September 30 of the same year, Senator Richard Blumenthal, chair of the subcommittee on consumer protection, product safety, and data security, convened a hearing titled "Protecting Kids Online: Facebook, Instagram, and Mental Health Harms" with only one witness: Facebook's global head of safety, Antigone Davis. Davis staunchly defended her company's work, arguing that the revelations made by Frances Haugen to the *Wall Street Journal* were a mischaracterization. She claimed that while their applications had some negative impacts, they were generally positive tools that actually helped teenagers navigate the challenges of their age. However, the interrogating members of Congress did not share this sugar-coated view. Senator Richard Blumenthal devastatingly declared that

> Facebook has taken big tobacco's playbook, it has hidden its own research on addiction and the toxic effects of its products. It has attempted to deceive the public and us in Congress about what it knows, and it has weaponized childhood vulnerabilities against children themselves. It has chosen growth over children's mental health and well-being, greed over preventing the suffering of children. These internal Facebook studies are filled with recommendations—recommendations from Facebook's own employees. And yet there is no evidence, none, that Facebook has done anything other than a few small, minor, marginal changes. We all know that Facebook treated protecting kids with disregard. If it had protected kids like it did drive up revenue or growth, it would have done a whole lot more. Instead, Facebook has evaded, misled and deceived.[24]

A few days later, on October 5, 2021, Mark Zuckerberg posted the following on Facebook: "At the heart of these accusations is this idea that we prioritize profit over safety and well-being. That's just not true."[25] The corporation's stance was clear: they were not sorry, they would not be held accountable, they would not facilitate access to the requested information, they did not accept responsibility, and they would not accept any sovereignty over Facebook other than their own. Haugen's testimony served to make visible and catalyze the plurality of critical voices against the power of big tech and the extensive data harms they have caused. And this was just one among numerous scandals and legal proceedings in which big tech companies were involved.

In the past five years alone, three companies, Amazon, Alphabet, and Meta, have accumulated over $15 billion in fines. The following are just a few examples. Meta has faced numerous sanctions for privacy violations, mishandling user data, and failing to comply with European court mandates. Significantly, in 2023, the Ireland's Data Protection Commission imposed a $1.3 billion fine on Meta for failing to comply with the European mandate to strengthen the security of the data transfers from European citizens to the United States.[26] This is a deeply interesting case that traces its origins back to a complaint filed in 2013 by a law student named Max Schrems before the Irish Data Protection Commissioner. Schrems, a digital rights activist, denounced Facebook's involvement in the US-organized mass surveillance program PRISM. Schrems demanded that European authorities halt Facebook's transfer of European data to the United States, a *harbor* no longer safe. It seemed like a quixotic fight, but in 2014, European courts ruled in favor of the activist.[27] In 2019, the US Federal Trade Commission imposed a $5 billion fine on Facebook, "the largest ever imposed on any company for violating consumers' privacy," as a result of the Cambridge Analytica scandal.[28]

Alphabet has faced enormous fines for its proven abuse of its dominant market position. Specifically, the EU fined Google €4.34 billion in 2018 for the restrictions imposed on Android device manufacturers.[29] In 2017, another fine of €2.42 billion was imposed for manipulation of search results.[30] Similarly, Amazon has incurred fines for anticompetitive behavior and data protection violations. The US Federal Trade Commission imposed a $30 million penalty for privacy violations of its Alexa and Ring users.[31] Notably, in 2021, the EU imposed a €746 million fine on Amazon for benefiting from its privileged position as a marketplace and seller. As demonstrated, Amazon fraudulently used data from its competing participants on the platform.[32] Additionally, France's Commission Nationale de l'Informatique et des Libertés (National Commission on Informatics and Liberty) imposed a €35 million fine on Amazon in the same year for the use of cookies without consent.[33] And this is only a fraction of the countless litigations in which big tech is involved.

Despite this storm of million-dollar sanctions, corporations have remained faithful to their criminal practices, highlighting the limitations of liberal responses to the social harm caused by their criminal behavior. In a fantastic article titled "What to do with the Harmful Corporation?," Marxist criminologist Steve Tombs presents what he believes (and I agree) are the problems with corporate sanctions.[34] First, he argues, the sanctions are not proportional to the social harm they cause; and, second, even if they were proportional in economic terms, they seem to be an inadequate method for dealing with corporate crime. Why? Even considering the potential high cost of sanctions, the chances of being inspected, detected, and truly pursued are reduced due to the structure of legal irresponsibility in which these corporate criminals operate. Third, the essentially punitive and reactive nature of sanctions does not foster a climate of corporate rehabilitation. After all, corporations, guided not by moral criteria but strictly by cost-benefit analysis, perceive the imposed sanctions as an externality, just another expense. Fourth, the impact of sanctions tends to dissipate as the costs are often dispersed among users and workers, for example, through salary reductions or price increases. Finally, the effects of sanction actions often fall exclusively on the corporation and not on the directors and executives who actually inspired the criminal behavior. Thus, the structure of corporate irresponsibility, built on the legal personality of the corporation and limited liability of the company, protects those who truly control and benefit from criminal conduct under the corporate veil. Nevertheless, the ineffectiveness of sanctions has led authorities in numerous countries to pursue alternative paths beyond mere sanctions: regulatory approaches that, just five years ago, were considered interventionist or even socialist. But is this enough?

Is It Possible to Regulate Big Tech?

In 2021, the then conservative Australian government forced Facebook-Meta to engage in a negotiation process with local media organizations. Despite the company's attempts at extortion, which led to the blocking of news on its platforms, it eventually agreed to the negotiation process.[35] That same year, in the United States, legal scholar Lina Khan, known for her critical work on Amazon's monopolistic behavior, was appointed as the Chair of the powerful Federal Trade Commission, responsible for ensuring market and consumer safety. Just a month after her appointment, Meta and Amazon formally requested Khan's recusal in cases involving these corporations, but their request was not granted. An indication of the Federal Trade Commission's interesting direction under Khan's leadership has been its attempt to "ramp up law enforcement against illegal repair restrictions," which has faced fierce resistance from powerful technology lobbies.[36] However, it is the European Union (EU) that is more decisively proposing comprehensive regulatory mechanisms to tackle the power of big tech.

The EU is implementing a political and legislative strategy aimed at providing a regulatory framework for the digital and technological spheres. The three fundamental instruments of the European strategy are the Digital Services Act, the Digital Markets Act, and legislation on Artificial Intelligence.[37] There are three main objectives of this regulatory effort. First, the European Union aims to establish a conducive political, economic, and legal framework to strengthen its technological sovereignty, upon which its economic, political, and social well-being also depends.[38] Second, linked closely to the first objective, the EU seeks to counteract the monopolistic control of foreign corporations. The European Commission has identified these corporations as a risk to fundamental rights, innovation, technological development, media plurality, and democratic values.[39] Finally, with the impetus of these regulations, the EU activates one of the distinctive features of its global soft power, known as the "Brussels effect."[40] Despite not being a military superpower or technological powerhouse comparable to China or the United States, the EU aims to exert global influence by defining norms, standards, and principles of technology governance. This approach has proven successful with the General Data Protection Regulation, which has de facto become the global standard for privacy legislation. Now, I will focus on the so-called Digital Services Act, which is the legislative instrument explicitly designed to regulate large digital platforms.

The Digital Services Act is specifically designed to address the threats posed by Very Large Online Platforms and Very Large Online Search Engines, which are defined as platforms with more than forty-five million monthly users. One of the most interesting features of the new regulatory framework is a shift in the focus of the protected legal interest. Traditionally, liberal legislations prioritize the individual and their rights while disregarding the structural and social dimensions in which the individual operates, often ignoring social, collective, or distributed harms. This was precisely one of the most criticized aspects of the General Data Protection Regulation and its excessive emphasis on individual privacy rights, a set of valuable but anachronistic rights in a society that is almost universally connected to the internet through proprietary technologies of major tech companies. The Digital Services Act recognizes the plurality of social, political, and economic interests threatened by large digital platforms:

> Given the importance of very large online platforms, due to their reach, in particular as expressed in the number of recipients of the service, in facilitating public debate, economic transactions and the dissemination to the public of information, opinions and ideas and in influencing how recipients obtain and communicate information online, it is necessary to impose specific obligations on the providers of those platforms, in addition to the obligations applicable to all online platforms. Due to their critical role in locating and making information retrievable online, it is also necessary to impose those obligations, to the extent they are applicable, on the providers of very large online search engines. Those additional obligations on providers of very large online platforms and of very large online search engines are necessary to address those public policy concerns, there being no alternative and less restrictive measures that would effectively achieve the same result.[41]

In contrast to previous models of online regulation, such as the German Network Enforcement Act, the new regulatory framework focuses primarily on governing the flow of information. To accomplish this, it proposes a set of targeted regulatory measures that encompass algorithmic recommendation systems, content moderation, illicit content management, and mechanisms for empowering users to uphold their fundamental digital rights. Moreover, the legislation reinforces users' control over their own data, aiming to enhance their agency in the digital realm. Equally significant, the legislation establishes a solid groundwork for ensuring transparency in the processes related to the governance of online discourse. This includes aspects like content removal, dissemination, and restriction. To guarantee the effectiveness of these transparency mechanisms, the legislation introduces the concept of independent audits, which serve as a means to verify compliance. Furthermore, in conjunction with the governance of online discourse, the legislation paves the way for the institutionalization of independent audits. This step acknowledges the importance of impartial evaluations in maintaining a fair and accountable digital environment. Notably, the legislation mandates the evaluation of algorithmic risks, encompassing concerns such as the dissemination of illegal content, potential adverse effects on fundamental rights, and the manipulation of services. By addressing these risks, the regulatory framework aims to safeguard the online ecosystem and protect the well-being of users.

In this sense, the new legislation requires large platforms to thoroughly assess the possible risks and social impacts in four major systemic categories. The first category refers to the dissemination of illicit content on digital platforms, with a focus on issues such as child abuse and hate speech. The second category addresses the potential impact of platforms on the exercise of fundamental rights recognized in European regulations, such as through the use of algorithmic tools that may lead to discrimination or censorship tools that result in the silencing of legitimate discourse. The third category emphasizes the threats that misinformation and manipulation can pose to civic and electoral processes. The fourth category highlights the potential impacts of digital platforms on the physical and mental well-being and safety of individuals.

The Digital Services Act establishes the creation of a European institutional apparatus for governing platforms. At the corporate level, large platforms will have to designate one or more persons who are responsible for enforcing European regulations and submitting periodic reports on their compliance.

However, the most interesting and relevant feature from the perspective of institutional governance of platforms is the creation of a network of digital services coordinators appointed at the national level by the member states. These coordinators will have broad powers of investigation, information, cessation, and sanction over digital platforms established within their jurisdictional scope. Nevertheless, it is the European Commission that holds the true power of investigation, inspection, and sanction over very large online platforms. The commission, acting on its own initiative, is endowed with severe punitive powers, including the ability to impose sanctions of up to 6 percent of the corporation's annual turnover. Undoubtedly, the new regulatory framework proposed by the European Commission represents a qualitative leap. The Digital Services Act echoes the vibrant academic and political discussions highlighting the impossibility of addressing data power without establishing governance mechanisms for large digital platforms. However, this model carries strong deficiencies from the previous model.

Limits of the Regulatory Efforts

Despite the advancements of the new legislation, the European Commission's proposal is not sufficiently equipped to deal with the crimes of digital capitalism. Large digital platforms are not reliable or suitable partners for democratic governments as the regulations presume; they are, in fact, excellent examples of corporate criminals.[42] These are companies that, in order to maintain their business model, do not hesitate to cause profound social harm.[43] The big platforms are not guided by the same moral and social paradigms of society. As the neoliberal guru Milton Friedman pointed out, they organize their actions under a logic of absolute market, and their only social responsibility is towards their shareholders.[44] Despite the vast differences in corporate culture, objectives, market, or platforms among companies like Alphabet, Meta, and Amazon, they all share a fundamental element. Their central business model is built on the control, exploitation, and monetization they exercise over vast sets of data. The short but impactful criminal history of big tech demonstrates two things: (1) the construction and maintenance of this model requires the systematic violation of individual and collective rights, as well as the democratic and "market" values that liberal governments claim to defend; and (2) despite countless legal proceedings and high-level investigations, corporations have proven to be incorrigible recidivists.

Further, as mentioned before, purely punitive power is not enough. The commission's threats, which amount to a significant 6 percent of annual turnover, fail to undermine the profits that large corporations obtain by maintaining their corporate criminal behavior. For example, Facebook-Meta and Google-Alphabet have been fined astronomical amounts totaling billions of dollars by US, UK, French, German, and European regional authorities, yet their criminal behavior has not ceased.[45] Furthermore, as I explained in chapter 2, lawmakers worldwide continue to entrust the management of sectors such as education or public administrations to well-known corporate criminals like Google-Alphabet.

Finally, another deficiency of the European Commission's proposal is the limited access that authorities and social organizations have to the algorithms that govern the flow of online information. Algorithms are part of the fundamental infrastructure of the digital era. They constitute the critical architecture of platform governance.[46] Without access to them, any "classic" paper-based law that seeks to regulate platforms will be doomed to failure. The new algorithmic landscape requires new legal instruments that, beyond the old efforts of legal codification, can translate democratic values into the algorithmic realm that European institutions claim to defend. Is it possible to do something about it? Can tools from critical criminology and abolitionism offer us some insights? I believe they can.

Navigating Contradictions

In the previous chapter, I made a conscious and critical defense of the need to criminalize ecocidal corporate behavior. My argument was not based on punitive morality—that all evil must be punished—but rather, it was situational and political. As I mentioned, and something I still maintain, it is necessary to use all available tools to stop the destruction and social harm caused by corporate criminals. But is this a solution? No, or at least I don't believe it is. In my opinion, it is merely a first step, among many others that need to be taken, which I will try to explain below. But before that, I want to focus on the necessary criticism raised by sectors of critical criminology and legal studies regarding corporate criminalization initiatives. I largely agree with the brilliant group of scholars from the new Marxist school of law, as well as with abolitionist criminology. The sharp criticism from these voices is particularly useful for scrutinizing the significant problems that arise when dealing with the punitive power of the state, as well as when attempting to regulate and contain capitalist institutions. Therefore, I believe it is necessary to present some of their arguments that highlight the weaknesses involved in advocating for the criminalization of behaviors, even when they are as socially harmful as those of the powerful.

A significant part of Marxist and abolitionist criminology agrees in pointing out that criminal law, including international law, is not an adequate tool for emancipation, as it reinforces the repressive and punitive power of the state. Historically, it has been directed against the oppressed. In any case, even if we were to accept the criminalization of some behaviors, it would be a repressive exercise destined to fail. As I mentioned in the previous chapter, there is a socio-legal culture that underestimates the importance and seriousness of the social harm caused by corporations, often opting for administrative offenses instead of crimes. Similarly, the bodies responsible for monitoring crimes committed by the powerful are chronically underfunded, lacking the personnel, training, and necessary technologies to address this type of criminality. Furthermore, as pointed out by Grietje Baars, David Whyte, Steven Bittle, Paddy Hillyard, and Steve Tombs, among others, the legal architecture in which corporations operate makes them particularly resistant to any attempt at criminalization.[47] The very structure of the criminal process makes it difficult, if not impossible, to prove the guilt, intentionality, and knowledge of legal persons. As has been noted in countless places, even if corporations are prosecuted, they are commonly offered the opportunity to reach settlements to mitigate the harm caused, settlements that often do not even cover a fraction of the material costs of their criminal behavior. In addition to these material limitations, there are notable ideological and political barriers.

Following the Marxist theory of alienation and, to a large extent, the critiques of Antonio Gramsci and George Luckacks, Grietje Baars argues that the domination of the ruling class over the oppressed is not only material but also ideological. International criminal law would be a pillar of this domination.[48] Robert Knox agrees with this critique of international law, which, following the anticolonial and Marxist school, is considered not just a product of imperialism but its juridified reflection.[49] Baars and David Whyte agree in pointing out that a potential court knowledgeable about international corporate crimes represents a symbolic false hope of the international community, the material outcome of which would be to strengthen the legitimacy of the system. Whyte provides the example of the International Criminal Court, stating that "the chances of ending up in the dock at the Hague are roughly zero unless you are African and unless you are on the wrong side of a Civil War. This is not just invective: in the first decade of the ICC's operation, only Africans had been brought to trial."[50] On the one hand, the criminalization of offenses such as ecocide would only allow the condemnation of a handful of acts considered particularly monstrous while absolving the criminal structure as a whole. On the other hand, the juridification of important political conflicts would hinder "real resistance to corporate power in the global political economy," paralyzing and institutionalizing the struggles. In another work, Whyte brilliantly summarizes the inherent contradictions in the demands for regulation and criminalization of corporate behavior. "By punishing the corporation," Whyte writes, "the system can claim it is intervening to protect the workers, the community, and so on, whilst at the same time maintain the steady rate of production, consumption, and financial transactions. We can call this a principle of regulatory tolerance, whereby the system upholds regulatory standards whilst at the same time tolerating corporate offending."[51]

Whyte's criticism extends to some proposals of the so-called corporate death penalty, raised among others by Mary Kreiner Ramirez. This mechanism has been theorized as an "optimized regime of punishment that includes termination of corporate existence in a systematic and rationalized manner" aimed at reimposing law and the rule of law.[52] The involuntary dissolution of a corporation could be determined as a consequence of its repeated and substantial violation of certain laws, a sort of "corporate three strikes," or as a result of the imposition of sanctions or civil liabilities high enough to result in the liquidation of the corporation. The problem with this approach, according to Whyte, is twofold. First, the responsibility of those who have truly been in a position of corporate power or who have benefited from the harmful acts caused by the corporation would be diluted behind the veil of corporate liquidation. Second, the "execution" of the corporation could have negative consequences for the most vulnerable and disadvantaged sectors, namely the workers and their communities.

However, despite the argumentation, Whyte, along with others like Tombs, does not advocate for standing idly by in the face of corporate crimes. One of the alternatives these authors defend (in addition to other political strategies and social mobilization) is the concept of equity fine. This is a sanctioning mechanism whereby shareholders who benefit from the corporation's activities are obliged to assume criminal responsibilities through the resocialization of part or all of their shares. These shares could be controlled, for example, by a social body designated by the state, the community, or even the workers themselves. In other words, instead of sentencing the corporation to death, this mechanism would allow for its restoration, its transformation into a new social entity controlled by the community. Additionally, and this is a shared vision among many social sectors, the authors call for an end to limited liability and, thereby, the responsibility of shareholders, owners, and other actors regarding the damages caused by actions they benefit from.

These authors are far from naive; they acknowledge that these mechanisms alone do not hold the potential for systemic change. They recognize that broader mobilizations and democratic transformations are necessary to bring about genuine systemic shifts. Despite their opposition to the existing bourgeois regulatory framework and the criminalization of corporate offenses, Whyte's and Tombs's proposals still require a legal process and some form of regulatory body to implement and enforce the measures. In other words, critical Marxist perspectives cannot currently renounce the legal form or the legal system, despite their opposition to it. This tension between opposition and necessity has been a consistent feature of historical revolutionary movements that genuinely strive for social transformation. While some may perceive this as an insurmountable paradox, particularly within intellectual circles, I view it as an explanation of the complex and dialectical nature of our current material conditions. As the former vice president of Bolivia, Álvaro García Linera, aptly stated, it reveals the inherent dialectical brutality embedded within our society: "To be in the State and simultaneously outside the State is a contradiction. But it is in riding this contradiction that the key to the continuity and defense of the progressive experience of democracy as a construction of equality lies."[53]

Marxist critiques have been part of my intellectual consciousness for a long time, and I still hold on to them to a great extent. As an abolitionist, I mistrust the punitive apparatus of the state. Prison is not, and will not be, the answer to any social problem. The process of securitization or the development of mega penal-punitive structures will also not serve to contain powerful historical dynamics of destruction and capital accumulation. As a pseudo-intellectual of the global loony left, I join the anticapitalist challenge that believes social transformation is impossible through mere reforms and legislative patchwork. In fact, I would say that corporations cannot be regulated. These behemoths are born to generate monetary wealth for their owners and investors, no matter the cost, and they will always find ways to render regulations redundant. But even so, and despite all its contradictions, I believe it is necessary to criminalize corporate behaviors such as mass privacy violations (as in the case of Cambridge Analytica) or ecocide (as in the case of Atacama).

In my opinion, legal reality is not defined by univocal truths and absolute powers. It is the result of a clash of forces, a tension, a contradiction. It is true that colonial capitalist powers have constructed a hegemonic order antagonistic to life. It is also true that almost all resistance has been annihilated in much of the global North. But the world is not limited to this minority. The Eurocentric paradigm that considered these privileged territories as the measuring stick for the rest of the world has long been discredited. The Global South offers numerous examples of resistance, of spaces of hope, and examples where hegemonic law is counterfeited by humble yet firm people's law. Perhaps it is because my roots lie in Latin America, or because I grew up there as an intellectual, but I believe, along with many others like Jesús de la Torre Rangel, that there is a law that can serve as a weapon of liberation.[54]

This belief is not an act of faith but of historical consciousness. It is necessary to reflect on the echoes of revolutionary legal history, ranging from the rural communities in Guerrero, where Indigenous people have been building law and autonomy for centuries, to la Havana,[55] where despite everything, even the forgetfulness of the Global North, socialism with its contradictions continues to be constructed. That law can be a tool for transformation is not something that has been said only in Latin America. All processes of decolonization, all struggles for emancipation and tyranny, have had their legal reflection. Law is a battlefield.[56] I think it is nice to think and believe in pure theories of revolution, but politics has been, is, and will be impure. In this sense, it can be said that while trusting and promoting punitive law reinforces the state, so does entrusting this same apparatus with the welfare systems that progressive positions generally defend. I am not going to be one who aligns with neoliberals, yippies, and conspiracy theorists attacking public health and education while defending the benevolence of private actors, be they pharmaceutical giants or Waldorf schools. On the contrary, my position is clear. I defend the right to health and universal education, which currently depend on a state apparatus for their provision, at least for now, at least in the Global North. Because if we look at other examples, the experience is different. Healthcare and education, not as disciplinary institutions but as mechanisms for the provision of care and communal knowledge, have existed in all communities and civilizations, regardless of if they had a state and regardless of whether they have suffered the consequences of capitalism. The same can be said of other institutions such as law and security, which, paraphrasing what García Linera said, are in the State and outside the State, riding the contradiction.[57]

What to Do Then?

At this point, we are faced with the eternal question of what to do in the face of the evils of capitalism. Or rather, as intelligently raised by the Tiqqun collective, how is this to be done? In this sense, I would like to conclude the book with a few brief lines, which are not programmatic but are, hopefully, a way to serve potential collective action.

All Fires Are the Same Fire

In my opinion, it is a privilege of the rich and the intellectuals to condemn certain forms of resistance. If the history of the oppressed teaches us something, it is that any site of struggle—the factory, the school, the family, the forest, the fields, life altogether—is a frontline, and hence everyone and every effort is necessary. In this sense, labeling one form of resistance as too radical or not conforming to what history demands is an exercise in arrogance, one that is typical of centralist universalism that we must unearth. Firstly, and I say this as a lawyer who has accompanied resistance movements, it is essential to fight against the offensive of digital capitalism with whatever tools we have at our disposal, no matter how modest or petty bourgeois they may appear. An example of this can be found in the General Data Protection Regulation. As Clifford and collegues pointed out in a recent article, despite its essentially focused nature on the defense of individual privacy rights, this legal instrument nevertheless provides tools for exercising collective resistance. This is the case with Article 22 and the defense it allows for workers' rights in digital contexts.[58] Similarly, as lawyers who accompany movements know well, bourgeois instruments offer tools that allow, if not a systemic change, at least the mitigattion or paralysis of the projected harm or ongoing illegal actions, for instance, as is the case with the projected lithium mining in Cáceres. Legal struggle and strategic litigation are not, and cannot be, an end in themselves, but they are a way to buy time to articulate long-term responses. Similarly, we should not be bound to the petty bourgeois false morality so inclined to condemn "mob's violence" while ignoring the structural apocalypses we all are immersed in. What is the best way to resist? The answer is not carved in stone; the ground, the people, and the time will tell.

Criminalization of the Powerful and Abolitionism Are Not Contradictory Terms

There is a false dilemma in theoretical discussions that stems from a misunderstanding of the purpose of criminalization. When I advocate for the criminalization of behaviors such as ecocide that jeopardize the material and social foundations of our societies, I am not endorsing the punitive power of the state. Instead, I am calling attention to grave offenses and demanding accountability. It is important to note that criminalization does not necessarily imply resorting to imprisonment as the only solution. There are alternative and innovative approaches that can arise from these processes, such as the concept of equity fines mentioned earlier. In this context, I believe that the role of critical abolitionist criminology goes beyond advocating for the abolition of prisons; it involves constructing new tools and institutions that render punitive measures obsolete and pave the way for transformative justice. Numerous abolitionist collectives are already engaged in thought-provoking work, exploring avenues for justice that are not only restorative but also transformative. Therefore, it is crucial for critical criminology to embrace the task of envisioning and developing the mechanisms that operationalize this transformation, especially when addressing corporate actors responsible for inflicting significant social harm. Rather than viewing criminalization solely as a means of punishment, we should seize the opportunity to reshape our understanding of justice. By doing so, we can move beyond the current system and create a more equitable and transformative model that addresses the root causes of harm while fostering healing and social change.

Only Political Response and Organization Will Allow Us to Combat the Ills of Digital Capitalism

If there is any agreement within the broad spectrum of leftist and progressive perspectives critical of capitalism, it is the belief, perhaps even a glimmer of hope, that the harms inflicted by this system of domination can only be effectively addressed through collective political action. In other words, the system will not change itself, and there are no magical solutions that will emerge out of thin air. It is our responsibility as the people to drive the necessary change. Currently, there are numerous initiatives and academic manifestos presenting varying degrees of utopian visions that guide us towards the prohibition of harmful and racist technologies, the dismantling of big tech hegemony, the establishment of technological sovereignty, and, on a more ambitious scale, the democratization of digital means of production. Personally, I had the privilege of being part of a diverse group of intellectuals and activists who generously contributed to the development of a programmatic framework for an emancipatory digital policy during the 2023 general elections in the Spanish state.

In summary, the program we devised encompasses the following key points:

1. Creation of a public agency for digital transition.
2. Public money, open-source code.
3. A public ecosystem of public-community platforms.
4. Guaranteeing the right to common ownership of data.
5. Creation of a national pact for technological innovation.
6. Sovereign national and pan-European technological infrastructures.
7. Ensuring democratic decision-making in technological development.
8. Digital downsizing in public administration.
9. Privacy and encryption of communications by design.
10. Network of facilities for digital inclusion.
11. Digital literacy with freely available educational materials and services promoted by the public administration.

12. Digital equality plans and promotion of feminist digitalization models.
13. An international order based on digital internationalism.

Plans like these are positive as they allow us to envision the horizons we want to strive for. However, plans are just plans. The fight against the crimes of digital capitalism will not be won through laws, plans, books, or institutions alone; they are merely tools. The battle against the crimes of digital capitalism, which is the battle against the powerful of our time, is being fought right now, in real time, by hundreds of thousands of people organized in labor unions, anti-racist collectives, hacker groups, political organizations, and communities resisting megaprojects; they are teachers, families, and friends. Only collective action, organized, conscious, joyful, and filled with love for life, will be capable of confronting this war that the powerful have been waging against the oppressed for centuries. Their hatred and disdain are so immense that they would let this world burn while fantasizing about escaping to Mars rather than stopping their greed. I have no trust, or interest in figures like Elon Musk, Mark Zuckerberg, or Jeffrey Bezos. However, I do have faith in that broad and diverse "us." Therefore, I will conclude with the words of Buenaventura Durruti, a revolutionary who fought and dreamed for the impossible: "We are not in the least afraid of ruins. We are going to inherit the earth. There is not the slightest doubt about that. The bourgeoisie might blast and ruin its own world before it leaves the stage of history. We carry a new world here, in our hearts. That world is growing in this minute."[59]

ACKNOWLEDGMENTS

Anyone who has ever stood in front of a blank page to write something like a book knows that what is produced is not the fleeting product of a brilliant genius. Words, ideas, are not individual gestures; they have roots, ligaments, ties, bonds, communities, but also debt, submission, and subjections. It is a collective process based on material conditions of privilege, friendship, affection, and conflict. It is necessary to make things clear. I wrote this book thanks to funding provided by the Australian Government to the University of Melbourne through the Australian Research Council ARC Centre of Excellence for Automated Decision-Making and Society (CE200100005). This money soon became a privilege. It became a visa for me and my family that many dream of and die for; without visas, many are imprisoned for "illegally crossing" the borders of a stolen land. That money became a salary that allowed us to peek into that mythical land of capitalist middle-class abundance. But I must be clear. That materiality was nourished by taxes derived from the plunder and pillage of a country where mining companies and large landowners set the pace of political power, the economy, and international relations. I am a direct beneficiary of colonization and dispossession, past and present. And for that I apologize. Specifically, this book was written in Melbourne, or Naarm. This is ravaged land, a land stolen from its guardians, the Wurundjeri people of the Kulin Nation, a land that still resists. To them, to their Elders past, present and future, I pay my respects and express my sincere gratitude. The colony will fall. It always was, always will be, Aboriginal land.

    I enjoyed the more than decent working conditions at the University of Melbourne. They were not the product of the benevolence of the powerful, but of the historic proletarian class struggle in so-called Australia. The decent conditions today are besieged by the neoliberal tempest and defended, as they were yesterday, by unions like the National Tertiary Education Union. I want to give thanks to the comrades who have been mobilizing for months; to those who went to the marches and picket lines; to those who organized and put up posters; and to those who taught in the streets and spoke about solidarity and resistance. Thanks especially to my colleagues from the Law School and the Criminology Department. Brilliant, good, and dignified people like Christine, Jake, Kate, Andrew, Megan, Fiona, Fan, Astari, Sara, Sahar, Juliet, and Laura. There are a number of radical researchers who, despite my punk attitude, have supported and believed in me trough the years. Thanks to Jule, Ekaitz, and James, who taught me that I was, indeed, a criminologist.

A city is not a city without the fabric of resistance running through it. In a life of nomadism, it is the rebellious spaces where I feel at home, comfortable. In these territories of life, there is always the promise, almost always fulfilled, of meeting wonderful people, people to learn and conspire with. Thanks to the critical nerds at the Institute of Postcolonial Studies, Carlos, Juan-Camilo, Lara, John, Scherezade, Tasnim, Eda, Nat, and especially to Jaz, heart, soul, and fire. Whatever good ideas there are in these pages come from them. Thanks to the chaotic, disparate, but always welcoming anarchist amalgam of the Catalyst and Black Spark. Whether training, reading, watching movies, or dining at a Food Not Bombs event, I have been able to get a little closer to the spark that will consume the chains of domination. Thanks to Tim, Jim, Emma, Jas, and Lol. I wish them the best as well as the end of capitalism. Finally, an important part of our lives in Naarm took place along the Brunswick East Primary School community, CERES, and the Merri Creek. I'm truly grateful our kid got to hang out in those awesome places with Hugo, Charlie, Flora, Emelyn, Willow, Walter, Isaac, Alma, and Tom (although they argued sometimes).

NOTES

INTRODUCTION

1. Gault 2021.
2. Soper, Tobin, and Smith 2021.
3. Orr et al. 2023.
4. Lata, Burdon, and Reddel 2023.
5. Rosenberg 2020.
6. Marinetti 2008.
7. Taplin 2017.
8. Gallagher and Carter 2023.
9. Liedke and Wang 2023.
10. Valtysson 2022.
11. Amazon 2022.
12. Kassem 2022.
13. Loewenstein 2023.
14. Loewenstein 2023.
15. Al Jazeera 2023.
16. Millman 2024.
17. Guterres 2023.
18. Malm 2016.
19. Griffin and Heede 2017.
20. Griffin and Heede 2017.
21. Supran, Rahmstorf, and Oreskes 2023.
22. J. Abraham 2016.
23. Joint Research Centre 2023.
24. European Commission 2023b.
25. Andreucci et al. 2023; Friends of the Earth Europe and Tansey 2023.
26. Cancela and Medina 2021.
27. Kerssens and van Dijck 2021.
28. Benvegnù et al. 2021, 697.
29. Richardson 2021.
30. Lipton 2020.
31. Engstrom et al. 2020.
32. United Nations 2019.
33. Richardson, Schultz, and Crawford 2019.

34 Wang 2018.
35 Jiménez 2020.
36 Klein 2020.
37 Andreessen 2011.
38 Soto Aliaga 2023; Gebrial 2022; Delfanti 2021b.
39 Cancela 2023.
40 Byler 2022.
41 Andreucci et al. 2023; Almeida et al. 2023.
42 Dyer-Witheford, Kjøsen, and Steinhoff 2019.
43 McQuillan 2022; Schaeffer 2022; Molnar 2024.
44 Andrejevic 2013.
45 Morozov 2011.
46 Couldry and Mejias 2020.
47 Tiqqun 2020.
48 Milmo and O'Caroll 2023.
49 Buck, n.d.; House of Lords 2021.
50 Allyn 2023.
51 Barrera and Bustamante 2018.
52 Horwitz, Seetharaman, and Wells 2021.
53 Horwitz, Seetharaman, and Wells 2021.
54 Pearce 1976.
55 Barak 2015, 106.
56 Senor and Singer 2011.
57 Pozo Marín and Benedicto 2022; Dana 2020.
58 Loewenstein 2023; Benedicto, Akkerman, and Brunet 2020.
59 European Commission 2019a.
60 Kilgore 2022; Wacquant 2009; Gilmore 2007.
61 Pashukanis 1983, 173.
62 Hillyard and Tombs 2007.
63 Gallagher 2023.
64 Kramer 1984.
65 Atiles-Osoria 2016.
66 Vitale 2021.
67 Jiménez and Douhaibi 2023; Richardson 2021.
68 Sutherland 1945.
69 Noble 2018; R. Benjamin 2019; and Broussard 2023.
70 Jefferson 2020.
71 Delio 2001.
72 Black 2012.
73 Muñiz 2022; Jefferson 2020.
74 Bedford et al. 2022.
75 Owen et al. 2022.
76 Pasternak et al. 2023; Spice 2018.

77 Riofrancos 2020.
78 Saner 2020.
79 Bedford et al. 2022.
80 Bisschop, Hendlin, and Jaspers 2022.
81 Whyte 2020.
82 Baars 2019.
83 Couldry and Mejias 2020.
84 Zuboff 2019; Andrejevic 2013.
85 Gebru 2020; Brayne 2020; Muñiz 2022; Molnar 2024.
86 Kukutai and Taylor 2016; Couldry and Mejias 2020.
87 Sadowski, Viljoen, and Whittaker 2021; Cancela 2023.

1. THE DIGITIZATION OF STATE RACISM

1   Heikkilä 2022.
2   Van Den Berg 2021.
3   Van Veen 2020.
4   Alston 2019a, 2019b.
5   Van Bekkum and Borgesius 2021.
6   Heikkilä 2022.
7   Broussard 2023.
8   Abrusci and Mackenzie-Gray Scott 2023; Green 2022.
9   Maki 2011.
10 Bekker 2021.
11 Vervloesem 2020.
12 Henley 2021.
13 Van Bekkum and Borgesius 2021.
14 Henley 2021.
15 Singh 2021.
16 Jiménez and Douhaibi 2023.
17 Whiteford 2021.
18 Eubanks 2018.
19 United Nations 2019.
20 Waldman 2019.
21 Richardson 2021, 795.
22 Hildebrandt 2018; Yeung 2018.
23 Jiménez and Douhaibi 2023; Chun 2021.
24 Mann 2020.
25 Braithwaite 2020.
26 Albanese 2023.
27 Karp 2023.
28 Commonwealth of Australia 2023, 31.
29 Zuberi 2001.
30 Castro-Gómez 2005.

31 Kalpagam 2014.
32 Rosenthal 2018.
33 Chun 2021.
34 Valdivia and Tazzioli 2023.
35 Hacking 1990.
36 Hacking 1990.
37 O'Neil 2016.
38 Noble 2018; Eubanks 2018; Chander and Krishnamurthy 2018; Morozov 2011.
39 Morozov 2021; Grewal and Purdy 2014; Harvey 2007.
40 Wacquant 2009.
41 A. Davis 2011, 2003; Gilmore 2022, 2007, 2002.
42 Gilmore 2007; Kundnani 2021.
43 Rodríguez 2020.
44 Moreton-Robinson 2015.
45 Quijano 2000; Grosfoguel 2016; Cusicanqui 2012b.
46 Quijano 2000.
47 C. Robinson 2020.
48 Horne 2014.
49 Simpson 2017; Coulthard 2014; Moreton-Robinson 2015; M. Davis 2022; Gilroy 2013.
50 Tauri and Porou 2014.
51 Porter 2016; Porter and Cunneen 2020.
52 Shilliam 2018.
53 Wang 2018.
54 R. Benjamin 2019.
55 Bonilla-Silva 2006.
56 Bonilla-Silva 2015.
57 Mezzadra and Neilson 2013.
58 Hampton 2021.
59 Mould 2020.
60 Gilliom 2001, 34.
61 Alexander 2010; Wacquant 2009; Gilmore 2007.
62 Eubanks 2018.
63 Urbán 2023; Brancoli 2023.
64 Weil 2009.
65 European Court of Justice 2017.
66 Hajjat and Mohammed 2023; García et al. 2021.
67 Kundnani 2014.
68 Tonnard 2011, 119.
69 United Nation 2019.
70 Pasquale 2020, 2015.
71 O'Neil 2016.
72 Crawford 2021.

73 Microsoft 2019.
74 Phan, Goldenfein, Mann and Kuch. 2022.
75 Australian Human Rights Commission 2020.
76 Appelman, Fathaigh, and Hoboken 2021.
77 Lighthouse Reports 2023.

2. THE DIGITAL TAKEOVER OF EDUCATION

1 Moore, Jayme, and Black 2021.
2 Kwet 2022; Cancela 2023.
3 Tarnoff 2022.
4 Dubal 2020.
5 Trades Union Congress 2021.
6 Daliri-Ngametua and Hardy 2022.
7 Feldman and Sandoval 2018.
8 Goldenfein and Griffin 2022.
9 Burrows 2012.
10 University of Queensland 2024.
11 Ovetz 2022, 2021.
12 Ovetz 2021.
13 Ovetz 2021.
14 Caplan 2009.
15 Caplan 2009.
16 Saunders 2010.
17 Stony Brook 2014.
18 Hern 2014.
19 Rushe 2013.
20 Patel 2013.
21 Electronic Frontier Foundation 2015.
22 Kastrenakes 2014.
23 Singer 2017.
24 Google 2020.
25 Google 2021.
26 NYU, n.d.
27 Chamayou 2021; Gago 2017; Foucault 2008.
28 McGettigan 2013.
29 Friedman 1962.
30 Connell 2013.
31 Muzzatti 2022; Collini 2017.
32 Hancox 2020.
33 Sorochan 2012.
34 Chamayou 2021.
35 Humber 2016; Mascarenhas 2012; Murray 2002.
36 Mulholland 2006.

37 Chakrabortty 2022.
38 Duggan et al. 2023.
39 Castillo 2020; Hursh and Martina 2003.
40 Saura 2016.
41 Hursh 2001, 355.
42 Hursh and Martina 2003.
43 Darling-Hammond 2007.
44 Prokop 2023.
45 Mateo, Ferrero, and Andrino 2022.
46 National Center for Education Statistics, n.d.
47 Holon IQ 2021.
48 Saura, Cancela, and Adell 2022.
49 Hamilton et al. 2022.
50 Lorenz 2012.
51 Grandinetti 2022.
52 Ivancheva and Garvey 2022.
53 Australian Capital Territory Government, n.d. b.
54 Australian Capital Territory Government, n.d. a.
55 Burnside Primary School 2022.
56 Warwick Valley Central School District 2021.
57 ACARA, n.d.
58 Northcote Primary School 2023.
59 Google 2022.
60 Google 2020.
61 Google, n.d.
62 Google Educator Groups Spain 2022.
63 Google Educator Groups Spain 2022.
64 Google Educator Groups Spain 2022.
65 Kumar et al. 2019; Lupton and Williamson 2017.
66 Yu and Couldry 2022.
67 Laird et al. 2022.
68 Ceres 2022.
69 Perrotta et al. 2021.
70 B. Robinson 2021.
71 Williamson 2017.
72 Datatilsynet 2022.
73 Williamson and Hogan 2020.
74 Krutka, Smits, and Willhelm 2021.
75 Collington 2022.
76 Pueyo Busquets 2022.
77 Xnet 2022.
78 New Mexico v. Google, pars. 4–5.
79 New Mexico v. Google, pars. 45–46.

80 New Mexico Department of Justice 2021.
81 Saura, Cancela, and Adell 2022.

3. CYBERWAR AGAINST THE PEOPLE

1. Y. Abraham 2023.
2. Goodfriend 2023; González 2022; Byler 2022.
3. Davies, McKernan, and Sabbagh 2023.
4. Sa'di 2021; Zureik 2020; Tawil-Souri 2012.
5. Zureik 2020.
6. Brooking and Campbell 2021.
7. Loewenstein 2023.
8. Molnar 2024; Beaumont 2022; Mijente, Just Futures Law, and No Border Wall Coalition 2021.
9. Schaeffer 2022.
10. Sterman 2022.
11. Farwell and Rohozinski 2011.
12. Trautman and Omerod 2017; Tiqqun 2020.
13. Arquilla and Ronfeldt 1997, 30.
14. Libicki 2009; Arquilla and Ronfeld 1993.
15. Muradov 2022; Robinson, Jones, and Janicke 2015.
16. Defence Australia 2019.
17. Haraway 2016, 7.
18. Churchill and Vander Wall 2002.
19. Lilli 2020.
20. Sanger 2017.
21. Molnar 2024, 2020.
22. Loewenstein 2023; Cristiano 2020; C. Miller 2018.
23. Schmitt 2013.
24. Tzouvala 2020.
25. Scholz and Galliott 2018.
26. Davidovic and Regan 2023; US Department of State 2023.
27. Leufer, Rodelli, and Hidvegi 2023.
28. Dyer-Witheford and Matviyenko 2019.
29. González 2022.
30. González 2022, 68.
31. Jefferson 2020.
32. Jefferson 2020.
33. Kilgore 2022.
34. McQuade 2019.
35. Neocleous 2014; Rigakos 2016.
36. McQuade 2019.
37. Muñiz 2022.
38. Zureik 2001, 2020.

39 European Council 2024.
40 International Organization for Migration 2022.
41 European Commission 2024b.
42 White House 2023.
43 Mijente, Just Futures Law, and No Border Wall Coalition 2021, 5.
44 Molnar 2020.
45 Salma 2022.
46 Roborder 2022.
47 European Commission 2024a.
48 Kassam 2022.
49 Mijente, Just Futures Law, and No Border Wall Coalition 2021.
50 T. Miller 2019.
51 Milivojevic 2022.
52 Mijente, Just Futures Law, and No Border Wall Coalition 2021.
53 Azek and Shah 2022.
54 Wang et al. 2022.
55 Phippen 2021.
56 State of California Department of Justice, n.d.
57 Muñiz 2022.
58 Kirchgaessner and Jones 2020.
59 Marzocchi and Mazzini 2022.
60 Scott-Railton et al. 2022.
61 Marzocchi and Mazzini 2022; Kaldani and Prokopets 2022; Scott-Railton et al. 2022.
62 UN Human Rights Council 2019.
63 Lyon 2015.
64 Greenwald and MacAskill 2013.
65 Greenwald and MacAskill 2013.
66 Privacy International 2013.
67 Van Dijck 2014.
68 Granick 2017.
69 Grothoff and Porup 2016.
70 Greenwald and MacAskill 2013.
71 Okpaleke et al. 2023.
72 Khan 2021.
73 Cachelin 2022.
74 Ahmed 2013.
75 Rotaru et al. 2022.
76 Rotaru et al. 2022, 1057.
77 Gilmore 2007.
78 US Census 2021.
79 Saunders, Hunt, and Hollywood 2016.
80 City of Chicago 2020.

81 Stroud 2021.
82 Kapustin et al. 2017.
83 Saunders et al. 2017.
84 Bertozzi, Brantingham, and Mohler 2014.
85 Bond-Graham and Winston 2014.
86 Munn 2018.
87 Stop LAPD Spying Coalition 2021.
88 Google Employees 2018.
89 Simonite 2021.
90 Haskins 2024.
91 Jewish Diaspora in Tech 2022.

4. THE MATERIALITY OF DIGITAL EXPLOITATION

1 Smalls, Maldonado, and Nieves 2022.
2 Zhou 2020; Transport Workers' Union 2020.
3 Orr et al. 2023.
4 Jiménez 2022.
5 Ball 2021; Nguyen 2021.
6 Alimahomed-Wilson and Reese 2021.
7 Stewart 2023.
8 Berardi 2003; Virno 2003.
9 Caffentzis 2013.
10 Lorey 2015.
11 Berry and McDaniel 2020.
12 Mezzadra and Neilson 2013.
13 Goikoetxea 2024.
14 Moreton-Robinson 2015.
15 Federici 2004.
16 Federici 2006, 7.
17 Jackson 2018; Simpson 2017; Coulthard 2014.
18 Marx 2004; Wang 2018.
19 Mejía Núñez 2022; Restrepo 2020.
20 Tankosić and Dovchin 2021.
21 Jayasuriya 2021; Reid, Ronda-Perez, and Schenker 2021.
22 Pasquinelli 2015; Delfanti 2021a.
23 Fuchs 2018.
24 Delfanti 2021b; Jamil 2020.
25 Tronti 2019, 12.
26 Marx and Engels 2010, 382.
27 Jiménez 2022.
28 Pitarch and Marco 2019.
29 Becker 2022; Lessig 2006.
30 Hassan and De Filippi 2017.

31 Cancela and Jiménez 2022; Schwarz 2019.
32 Transport Workers' Union 2020.
33 Gebrial 2022.
34 Prat and Ranz 2020.
35 Amazon 2022.
36 Mawhinney, Reinhard, and Lefebvre 2023.
37 Transport Workers' Union 2020.
38 Gebrial 2022.
39 Uber Eats 2023.
40 Uber 2018.
41 Uber 2022.
42 Barratt, Goods, and Veen 2023; Jiménez 2022.
43 Dubal 2022.
44 Conger 2021.
45 Adams-Prassl 2023; Álvarez Barba 2023; Hummel 2023.
46 Orr et al. 2023; Jamil 2020.
47 Dubal 2020.
48 Marx 2004, 694.
49 Dubal 2017.
50 Gebrial 2022.
51 Gebrial 2022.
52 Jamil 2020.
53 Diab 2019.
54 Chan 2019.
55 Jamil 2020.
56 Mawhinney, Reinhard, and Lefebvre 2023; Cram et al. 2022.
57 Orr et al. 2023.
58 Stewart 2022.
59 Trades Union Congress 2020.
60 Blanc and Maldonado 2022.
61 Smalls, Maldonado, and Valentin 2022.
62 Stewart 2023.
63 Gig Economy Project and Soto 2021.

5. LAW AND EXTRACTIVISM
1 Lithium Iberia, n.d. a.
2 Lithium Iberia, n.d. b.
3 European Parliament 2023a.
4 Junta de Extremadura 2022.
5 Pintos Cubo 2022.
6 Pérez Gómez 2021.
7 Lithium Iberia 2022.
8 Infinity Lithium, n.d.

9   Domínguez 2021.
10 Gómez 2023.
11 Gudynas 2021; Scott 2021; Acosta 2013.
12 elDiarioAr 2023.
13 Castelos 2023; Dunlap and Riquito 2023.
14 Spice 2018.
15 Djukanovic 2022.
16 Pasternak et al. 2023.
17 In Pasternak et al. 2023, 1.
18 Cowen 2020.
19 Watson 2018; Moreton-Robinson 2015.
20 Pasternak et al. 2023, 3.
21 Walton et al. 2021.
22 Beiser 2019.
23 Sibelco 2023.
24 Whyte 2014.
25 Wang and Tomaney 2019.
26 Bruno 2022.
27 Ned Nemra 2021.
28 Ng 2019.
29 Bisschop, Hendlin, and Jaspers 2022.
30 Cellular Telecommunications and Internet Association 2023.
31 CommScope, n.d.
32 Olivo 2023.
33 Google 2022.
34 Rogoway 2023.
35 Mytton 2021.
36 Li et al. 2023.
37 Mytton 2021.
38 Ward 2022.
39 Amazon 2023, 2.
40 Apple 2023, 6.
41 Uber 2020.
42 Amazon 2022.
43 Hund et al. 2020.
44 Fernández, González, and Ramiro 2022.
45 Whyte 2020.
46 Fernández et al. 2022.
47 Carroll 2020.
48 W. Benjamin 1996, 288.
49 Gorz 1994.
50 Schmelzer, Vetter, and Vansintjan 2022.
51 Biden 2023.

52. Von der Leyen 2022.
53. Jassy 2022.
54. Rio Tinto, n.d.
55. Rio Tinto 2022.
56. Carrington 2020.
57. Oberle et al. 2019.
58. United Nations 2019.
59. Global Foodprint Network 2023.
60. Global Forest Watch 2023.
61. Gonzaga 2022.
62. Almond et al. 2022.
63. Intergovernmental Panel on Climate Change 2021.
64. Von der Leyen 2021a.
65. International Monetary Fund 2021.
66. Muench et al. 2022.
67. Marr 2022.
68. Rio Tinto 2021.
69. Muench et al. 2022, iv.
70. Muench et al. 2022, v.
71. Joint Research Centre 2022.
72. European Round Table for Industry 2022.
73. Lasslett 2014.
74. Inuarak 2022.
75. LobbyFacts 2024.
76. Cancela and Medina 2021.
77. Lechanteaux 2021.
78. Ernst & Young 2021.
79. Bobba et al. 2020.
80. European Commission 2019b.
81. European Commission 2020.
82. Watson 2018.
83. UNECE 2022.
84. Von der Leyen 2022.
85. Digital Europe 2022.
86. Friends of the Earth Europe and Tansey 2023.
87. Breton 2022.
88. Almeida et al. 2023.
89. European Commission 2023b.
90. European Commission 2023a, 8.
91. Vidalou 2017; Troupe 2018; Del Mármol and Vaccaro 2020.

6. KILLING THE SALAR DE ATACAMA

1. Crook and Short 2014.
2. Zierler 2011.
3. Extinction Rebellion 2019.
4. Stop Ecocide 2021.
5. Watego 2021.
6. Atiles-Osoria 2014; Goyes 2019; Cunneen and Tauri 2016; Weis 2019.
7. Wolfe 2006.
8. Short and Crook 2022.
9. Spice 2018.
10. Simpson 2016, 23.
11. Sainato 2023.
12. The Basque Country, "land of the Basque language."
13. Large forest trees with prickly, olive-green leaves and reddish-brown bark that peels in long strips.
14. A green, copper, and white native bush pigeon.
15. A dark bluish-gray, rare forest bird of limited flight with a black facial mask, blue wattles, a short, strongly arched bill, long black legs, and a long tail.
16. A large native forest parrot with olive-brown and dull green upperparts and crimson underparts.
17. Red pine—a tall coniferous tree with dark brown flaking bark, scale-like prickly leaves, and gracefully weeping branches.
18. Celery pine—a tall forest tree with long, fan-like and leathery leaves which look like celery leaves.
19. Black pine—a coniferous, long-lived native tree of lowland forest with small, narrow leaves arranged in two rows and a hammer-marked trunk.
20. Palm-leaf fern—a robust native creeping ground fern with long drooping fronds.
21. Native tree fern with very thick, soft, fibrous, rusty-brown trunk and a heavy skirt of dead, pale-brown fronds.
22. Canoe.
23. Wilmshurst 2007; Knight 2009.
24. Extended kinship group, tribe—often refers to a large group of people descended from a common ancestor and associated with a distinct territory.
25. Chief (gender neutral).
26. Self-governance.
27. Māori language.
28. Custodianship, care.
29. Higgins 2018; Te Awekotuku and Nikora 2003; Binney 2009.
30. Food, meal.
31. Coulthard 2014; Simpson 2017.
32. King-Jones and Wright 2011.
33. Mankelow 2014.

34 Higgins 2018.
35 Prestige, authority, control, power, influence, status, spiritual power, charisma—*mana* is a supernatural force in a person, place, or object.
36 Life principle/force, vital essence, special nature, source of emotions—the essential quality and vitality of a being or entity.
37 Title to land through occupation by a group, generally over a long period of time.
38 Local people, Indigenous people.
39 Guardian, custodian.
40 Department of Conservation 2014, 8.
41 Geddis and Ruru 2019.
42 Cusicanqui 2012a.
43 Linera and Stefanoni 2008.
44 Wolkmer 2019.
45 Montoya, Sieder, and Bravo-Espinosa 2022.
46 Wolkmer 2019; Tapia 2007.
47 Schavelzon 2012.
48 Wolkmer 2019.
49 Horne 2014.
50 Gargarella 2013.
51 Rajagopal 2003.
52 Coulthard 2014; Simpson 2017, 2016.
53 Haines and Parker 2017.
54 Shemitz and Anastas 2020.
55 Benedicto, Akkerman, and Brunet 2020.
56 Watson et al. 2020.
57 Kirchman 2021.
58 Harper 2005.
59 Pellegrini et al. 2020.
60 Heal et al. 2019.
61 French and Kotzé 2019; UN Environment Programme 2018; Nellemann et al. 2016.
62 UN Environment Programme 2018; Fajardo del Castillo 2015.
63 UNEP-Interpol 2016, 30.
64 Europol 2022.
65 European Parliament 2023b.
66 Eurojust 2021.
67 European Commission 2021.
68 Stop Ecocide 2021.
69 Romano 2023.
70 US Geological Survey 2023.
71 European Commission 2023b; International Energy Agency 2023.
72 Goldman Sachs 2023.
73 International Energy Agency 2021.

74 Acosta 2013; Bárcenas and Eslava Galicia 2011.
75 Riofrancos 2020.
76 Hernandez and Newell 2022; Svampa 2019.
77 Escobar 2011; Alimonda 2019.
78 Mining Technology 2024.
79 Galaz-Mandakovic, Araya, and Rivera 2023.
80 Environmental Justice Atlas 2021.
81 Jerez, Garcés, and Torres 2021.
82 Dube 2022.
83 World Integrated Trade Solution 2021.
84 Katwala 2022.
85 Lerner 2012.
86 Briones and Delrio 2007; Alimonda and Ferguson 2004.
87 Moreton-Robinson 2019; Fitzmaurice 2007.
88 Gajardo and Redón 2019.
89 Vilca et al. 2022.
90 Bolados García and Babidge 2017.
91 Acuña and Tironi 2022.
92 Morales 2013.
93 Aylwin, Didier, and Mora 2021.
94 Bustos-Gallardo, Bridge, and Prieto 2021.
95 Jerez, Garcés, and Torres 2021.
96 Bustos-Gallardo, Bridge, and Prieto 2021.
97 Liu and Agusdinata 2020.
98 Albemarle, n.d.
99 Gajardo and Redón 2019.
100 Gutiérrez et al. 2022.
101 Ministerio de Agricultura 2023.
102 Liu and Agusdinata 2020.
103 Bolados García and Babidge 2017.
104 Dunlap 2021; Crook, Short, and South 2018.
105 Romero, Aylwin, and Didier 2021.
106 Aylwin, Didier, and Mora 2021.
107 United Nations 2007, 37.
108 United Nations 2007, 39.
109 Superintendencia de Medio Ambiente 2022.
110 European Commission 2023b.

CONCLUSION
  1 World Bank Data, n.d.
  2 Metz, Lewis, and Slimak 2023.
  3 Colledge et al. 2019.
  4 Brügger 2015.

5   Meta 2022.
6   Mumford 2010.
7   Rossini et al. 2021.
8   Kwet 2019.
9   Couldry and Mejias 2023.
10 Nemer 2018.
11 Evangelista and Bruno 2019.
12 Matamoros-Fernández 2017; Siapera and Viejo-Otero 2021; Kalsnes and Ihlebæk 2021.
13 Yue 2020; Venier 2019.
14 Warofka 2018.
15 Milmo 2021.
16 Federal Trade Commission 2019b.
17 Davies et al. 2022.
18 Jiménez and Oleson 2022.
19 Horwitz 2021.
20 Haugen 2021b; Haugen 2021a.
21 Redden, Brand, and Terzieva 2020.
22 Haidt 2021.
23 Haugen 2021b.
24 Blumenthal 2021.
25 Zuckerberg 2021.
26 European Data Protection Board 2023.
27 Jiménez and Oleson 2022; Jougleux 2022.
28 Federal Trade Commission 2019a.
29 European Commission 2018.
30 European Commission 2017.
31 Archie 2023.
32 Burgess 2021.
33 CNIL 2022.
34 Tombs 2016.
35 Bossio et al. 2022.
36 Federal Trade Commission 2021.
37 Jiménez and Cancela 2023.
38 Cancela and Jiménez 2022.
39 Klossa 2019.
40 Bradford 2020.
41 Regulation (EU) 2022/2065 (Digital Services Act), recital 75.
42 Tombs and Whyte 2015.
43 Canning and Tombs 2021.
44 Friedman 1970.
45 Jiménez and Oleson 2022.
46 Gorwa 2019.

47 Baars 2019; Whyte 2020; Bittle 2012; Hillyard and Tombs 2007.
48 Baars 2019.
49 Knox 2023.
50 Whyte 2020, 162.
51 Whyte 2019, 294.
52 Ramirez 2005.
53 Linera 2022.
54 De la Torre Rangel 2006.
55 Sierra 2017; Burgos Matamoros 2017.
56 de Cabo Martín 2022; Bourdieu 1987.
57 Linera 2022.
58 Clifford et al. 2023.
59 Buenaventura Durruti, as quoted in Enzesberger 2020, 157.

BIBLIOGRAPHY

Abraham, John. 2016. "Worst Mediterranean Drought in 900 Years Has Human Fingerprints All Over It." *The Guardian*, March 18, 2016. https://www.theguardian.com.
Abraham, Yuval. 2023. "A Mass Assassination Factory: Inside Israel's Calculated Bombing of Gaza." *972 Magazine*, November 30, 2023.
Abrusci, Elena, and Richard Mackenzie-Gray Scott. 2023. "The Questionable Necessity of a New Human Right against Being Subject to Automated Decision-Making." *International Journal of Law and Information Technology* 31 (2): 114–43.
Acosta, Alberto. 2013. "Extractivism and Neoextractivism: Two Sides of the Same Curse." In *Beyond Development: Alternative Visions from Latin America*, edited by Miriam Lang and Dunia Mokrani, 61–86. Amsterdam: Transnational Institute.
Acuña, Valentina, and Manuel Tironi. 2022. "Extractivist Droughts: Indigenous Hydrosocial Endurance in Quillagua, Chile." *Extractive Industries and Society* 9 (March): 101027.
Adams-Prassl, Jeremias. 2023. "*Uber BV v Aslam*: '[W]ork relations . . . Cannot Safely Be Left to Contractual Regulation.'" *Industrial Law Journal* 51 (4): 955–66.
Ahmed, Akbar. 2013. *The Thistle and the Drone: How America's War on Terror Became a Global War on Tribal Islam*. Washington DC: Brookings Institution Press.
Al Jazeera. 2023. "Elon Musk Meets Netanyahu during Israel Visit." *Al Jazeera*, November 27, 2023. https://www.aljazeera.com.
Albanese, Anthony. 2023. "Press Conference—Canberra." Transcript. July 7, 2023. https://www.pm.gov.au.
Albemarle. n.d. "Lithium Resources." Accessed April 14, 2023. https://www.albemarle.com.
Alexander, Michelle. 2010. *The New Jim Crow: Mass Incarceration in the Age of Colorblindness*. New York: New Press.
Alimahomed-Wilson, Jacob, and Ellen Reese. 2021. "Surveilling Amazon's Warehouse Workers: Racism, Retaliation, and Worker Resistance amid the Pandemic." *Work in the Global Economy* 1 (1–2): 55–73.
Alimonda, Héctor. 2019. "The Coloniality of Nature: An Approach to Latin American Political Ecology." *Alternautas* 6 (1): 102–42.
Alimonda, Héctor, and Juan Ferguson. 2004. "La Producción del desierto: Las imágenes de la campaña del Ejército Argentino contra los indios, 1879." *Revista Chilena de Antropología Visual*, no. 4, 1–28.
Allyn, Bobby. 2023. "States Sue Meta, Claiming Instagram, Facebook Fueled Youth Mental Health Crisis." *NPR*, October 24, 2023. https://www.npr.org.

Almeida, Diana Vela, Vijay Kolinjivadi, Tomaso Ferrando, Brototi Roy, Héctor Herrera, Marcela Vecchione Gonçalves, and Gert Van Hecken. 2023. "The 'Greening' Of Empire: The European Green Deal as the EU First Agenda." *Political Geography* 105 (August): 102925.

Almond, Rosamunde E. A., Monique Grooten, Diego Juffe Bignoli, and Tanya Petersen. 2022. *Living Planet Report 2022: Building a Nature-Positive Society*. Gland, Switzerland: World Wildlife Fund.

Alston, Philip. 2019a. *Report of the Special Rapporteur on Extreme Poverty and Human Rights*. New York: United Nations.

Alston, Philip. 2019b. "Brief by the United Nations Special Rapporteur on Extreme Poverty and Human Rights as *Amicus Curiae* in the Case of NJCM c.s./De Staat der Nederlanden (SyRI) before the District Court of The Hague (case number: C/09/550982/HA ZA 18/388)." OHCHR, accessed January 31, 2024. https://www.ohchr.org.

Álvarez Barba, Yago. 2023. "Un año de la Ley Rider y Uber quiere dar un paso atrás: ¿Qué está fallando?" *El Salto*, August 13, 2023. https://www.elsaltodiario.com.

Amazon. 2022. *Delivering Progress Every Day: Amazon's 2021 Sustainability Report*. N.p.: Amazon. https://sustainability.aboutamazon.com.

Amazon. 2023. *Building a Better Future tTogether: Amazon Sustainability Report 2022*. N.p.: Amazon. https://sustainability.aboutamazon.com.

Andreessen, Marc. 2011. "Why Software Is Eating the World." *Wall Street Journal*, August 20, 2011.

Andrejevic, Mark. 2013 "Surveillance in the Digital Enclosure." In *The New Media of Surveillance*, edited by Shoshana Magnet and Kelly Gates, 18–40. London: Routledge.

Andreucci, Diego, Gustavo García López, Isabella M. Radhuber, Marta Conde, Daniel M. Voskoboynik, J. D. Farrugia, and Christos Zografos. 2023. "The Coloniality of Green Extractivism: Unearthing Decarbonisation by Dispossession through the Case of Nickel." *Political Geography* 107:102997.

Appelman, Naomi, Ronan Ó. Fathaigh, and Joris van Hoboken. 2021. "Social Welfare, Risk Profiling and Fundamental Rights: The Case of SyRI in the Netherlands." *Journal of Intellectual Property, Information Technology and E-Commerce Law* 12 (4): 257.

Apple. 2023. *Environmental Progress Report: Covering Fiscal Year 2022*. N.p.: Apple Inc. https://www.apple.com.

Archie, Ayana. 2023. "Amazon Must Pay Over $30 Million Over Claims It Invaded Privacy with Ring and Alexa." *NPR*, June 1, 2023. https://www.npr.org.

Arquilla, John, and David Ronfeldt. 1993. "Cyberwar is Coming!" *Comparative Strategy* 12 (2): 141–65.

Arquilla, John, and David Ronfeldt. 1997. *In Athena's Camp: Preparing for Conflict in the Information Age*. Santa Monica: RAND.

Atiles-Osoria, José Manuel. 2014. "The Criminalization of Socio-environmental Struggles in Puerto Rico." *Oñati Socio-Legal Series* 4 (1): 85–103.

Atiles-Osoria, José Manuel. 2016. "Colonial State Terror in Puerto Rico: A Research Agenda." *State Crime Journal* 5 (2): 220–41.

Australian Capital Territory Government. n.d. a. "Future of Education—Digital Access and Equity." Australian Capital Territory Government: Education. Accessed May 23, 2023. https://www.education.act.gov.au.

Australian Capital Territory Government. n.d. b. "Future of Education—Digital Access and Equity Program: Frequently Asked Questions." Australian Capital Territory Government: Education. Accessed June 19, 2024. https://www.education.act.gov.au.

Australian Curriculum, Assessment and Reporting Authority (ACARA). n.d. "Information and Communication Technology (ICT): Capability (Version 8.4)." Accesed March 12, 2023. https://www.australiancurriculum.edu.au.

Australian Human Rights Commission. 2020. *Using Artificial Intelligence to Make Decisions: Addressing the Problem of Algorithmic Bias*. Sydney, Australia: Australian Human Rights Commission.

Aylwin, José, Marcel Didier, and Oriana Mora. 2021. *Evaluación de impacto en derechos humanos de SQM en los derechos del pueblo indígena Lickanantay*. Santiago, Chile: Observatorio Ciudadano.

Azek, Mizue, and Paromita Shah. 2022. *Hart Attack: How DHS's Massive Biometrics Database Will Supercharge Surveillance and Threaten Rights*. N.p.: Immigrant Defense Project, Mijente, Just Futures Law. https://surveillanceresistancelab.org.

Baars, Grietje. 2019. *The Corporation, Law and Capitalism: A Radical Perspective on the Role of Law in the Global Political Economy*. Leiden: Brill.

Ball, Kirstie. 2021. *Electronic Monitoring and Surveillance in the Workplace*. European Commission, Joint Research Centre. Luxembourg: Publications Office of the European Union.

Barak, Gregg. 2015. "The Crimes of the Powerful and the Globalization of Crime/Os Crimes dos Poderosos e a Globalização do Crime." *Revista Brasileira de Direito* 11 (2): 104–14.

Bárcenas, Francisco López, and Mayra Montserrat Eslava Galicia. 2011. *El mineral o la vida: La legislación minera en México*. Oaxaca: Centro de Orientación y Asesoría a Pueblos Indígenas.

Barratt, Tom, Caleb Goods, and Alex Veen. 2023. "Australia: Labour and the Gig Economy." In *The Routledge Handbook of the Gig Economy* edited by Immanuel Ness, 347–58. London: Routledge.

Barrera, Rita, and Jessica Bustamante. 2018. "The Rotten Apple: Tax Avoidance in Ireland." *International Trade Journal* 32 (1): 150–61.

Beaumont, Hilary. 2022. "'Never Sleeps, Never Even Blinks': The Hi-Tech Anduril Towers Spreading along the US border." *The Guardian*, September 16, 2022. https://www.theguardian.com.

Becker, Katrin. 2022. "Blockchain Matters—Lex Cryptographia and the Displacement of Legal Symbolics and Imaginaries." *Law Critique* 33 (2): 113–30.

Bedford, Laura, Monique Mann, Marcus Foth, and Reece Walters. 2022. "A Post-Capitalocentric Critique of Digital Technology and Environmental Harm: New Directions at the Intersection of Digital and Green Criminology." *International Journal for Crime, Justice and Social Democracy* 11 (1): 167–81.

Beiser, Vince. 2019. *The World in a Grain: The Story of Sand and How It Transformed Civilization.* New York: Penguin.

Bekker, Sonja. 2021 "Fundamental Rights in Digital Welfare States: The Case of SyRI in the Netherlands." In *Netherlands Yearbook of International Law 2019: Yearbooks in International Law: History, Function and Future,* edited by Otto Spijkers, Wouter G. Werner, Ramses A. Wessel, 289–307. The Hague: T. M. C. Asser Press.

Benedicto, Ainhoa Ruiz, Mark Akkerman, and Pere Brunet. 2020. *A Walled World: Towards A Global Apartheid.* Barcelona: Centre Delàs d'Estudis per la Pau. https://www.borderline-europe.de.

Benjamin, Ruha. 2019. *Race After Technology: Abolitionist Tools for the New Jim Code.* New York: Polity Press.

Benjamin, Walter. 1996. *Selected Writings: 1913–1926. Vol. 1.* Cambridge, MA: Harvard University Press.

Benvegnù, Carlotta, Niccolò Cuppini, Mattia Frapporti, Floriano Milesi, Maurilio Pirone, and (Into the Black Box). 2021. "Platform Battlefield: Digital Infrastructures in Capitalism 4.0." *South Atlantic Quarterly* 120 (4): 689–702.

Berardi, Franco. 2003. *La fábrica de la infelicidad. nuevas formas de trabajo y movimiento global.* Madrid: Traficantes de Sueños.

Berry, Craig, and Sean McDaniel. 2020. "Young People and the Post-Crisis Precarity: The Abnormality of the 'New Normal.'" *LSE British Policy and Politics* (blog), January 20, 2020. https://blogs.lse.ac.uk.

Bertozzi, Andrea L., P. Jeffrey Brantingham, and George Mohler. 2014. *Final Report: Dynamic Models of Insurgent Activity.* Los Angelos: University of California Los Angeles, Office of Contract and Grant Administration.

Biden, Joe. 2023. "Remarks by President Biden on Economic Progress since Taking Office." *White House,* January 26, 2023. https://www.whitehouse.gov.

Binney, Judith. 2009. *Encircled lands: Te Urewera, 1820–1921.* Wellington, New Zealand: Bridget Williams Books.

Bisschop, Lieselot, Yogi Hendlin, and Jelle Jaspers. 2022. "Designed to Break: Planned Obsolescence as Corporate Environmental Crime." *Crime, Law and Social Change* 78 (3): 271–93.

Bittle, Steven. 2012. *Still Dying for a Living: Corporate Criminal Liability after the Westray Mine Disaster.* Toronto: University of British Columbia Press.

Black, Edwin. 2012. *IBM and the Holocaust: The Strategic Alliance between Nazi Germany and America's Most Powerful Corporation-Expanded Edition.* New York: Dialog.

Blanc, Eric, and Angelika Maldonado. 2022. "Here's How We Beat Amazon." *Jacobin,* April 2, 2022. https://jacobin.com.

Blumenthal, Richard. 2021. Hearing before the Subcomm. on Consumer Prot., Product Safety & Data Sec. of the S. Comm. on Com., Sci. & Transp., 117th Cong. (Sept. 30. 2021) (statement of Sen. Richard Blumenthal, Chairman, Subcomm. on Consumer Prot., Product Safety & Data Sec.).

Bobba, Silvia, Samuel Carrara, Jaco Huisman, Fabrice Mathieux, and Claudiu Pavel. 2020. *Critical Raw Materials for Strategic Technologies and Sectors in the EU: A Foresight Study*. European Commission, Joint Research Centre. Luxembourg: Publications Office of the European Union.

Bolados García, Paola, and Sally Babidge. 2017. "Ritualidad y Extractivismo: la Limpia de Canales y las Disputas por el Agua en el Salar de Atacama-Norte de Chile." *Estudios Atacameños*, no. 54, 201–16.

Bond-Graham, Darwin, and Ali Winston. 2014. "From Fallujah to the San Fernando Valley, Police Use Analytics to Target 'High Crime' Areas." *Truthout*, March 12, 2014.

Bonilla-Silva, Eduardo. 2006. *Racism without Racists: Color-Blind Racism and the Persistence of Racial Inequality in the United States*. Washington, DC: Rowman & Littlefield.

Bonilla-Silva, Eduardo. 2015. "The Structure of Racism in Color-Blind, 'Post-Racial' America." *American Behavioral Scientist* 59 (11): 1358–76.

Bossio, Diana, Terry Flew, James Meese, Tama Leaver, and Belinda Barnet. 2022. "Australia's News Media Bargaining Code and the Global Turn towards Platform Regulation." *Policy & Internet* 14 (1): 136–50.

Bourdieu, Pierre. 1987. "The Force of Law: Toward a Sociology of the Juridical Field." *Hastings Law Journal* 38 (5): 814–53.

Bradford, Anu. 2020. *The Brussels Effect: How the European Union Rules the World*. Oxford: Oxford University Press.

Braithwaite, Valerie. 2020. "Beyond the Bubble That is Robodebt: How Governments that Lose Integrity Threaten Democracy." *Australian Journal of Social Issues* 55 (3): 242–59.

Brancoli, Fernando. 2023. *Bolsonarismo: The Global Origins and Future of Brazil's Far Right*. New Brunswick, NJ: Rutgers University Press.

Brayne, Sarah. 2020. *Predict and Surveil: Data, Discretion, and the Future of Policing*. Oxford: Oxford University Press.

Breton, Thierry. 2022. "Critical Raw Materials Act: Securing the New Gas & Oil at the Heart of Our Economy—Blog of Commissioner Thierry Breton." European Commission, September 14, 2022. https://ec.europa.eu.

Briones, Claudia, and Walter Delrio. 2007. "La 'Conquista del Desierto' desde perspectivas hegemónicas y subalternas." *Runa* 27 (1): 23–48.

Brooking, Emerson, and Eliza Campbell. 2021. "How to End Israel's Digital Occupation." *Foreign Policy*, December 3, 2021. https://foreignpolicy.com.

Broussard, Meredith. 2023. *More than a Glitch: Confronting Race, Gender, and Ability Bias in Tech*. Cambridge, MA: MIT Press.

Brügger, Niels. 2015. "A Brief History of Facebook as a Media Text: The Development of an Empty Structure." *First Monday* 20 (5).

Bruno, Margherita. 2022. "Ever Alot Breaks Record for World's Largest Containership." *Port Technology*, June 27, 2022. https://www.porttechnology.org.

Buck, Ken. n.d. *The Third Way: Antitrust Enforcement in Big Tech*. Draft of report for the House Judiciary Committee Subcommittee on Antitrust, Commercial, and Administrative Law. www.politico.com.

Burgess, Matt. 2021. "Why Amazon's £636m GDPR Fine Really Matters." *Wired*, August 4, 2021. www.wired.com.

Burgos Matamoros, Mylai. 2017. "Análisis crítico constitucional de los cambios sociojurídicos en la Cuba actual / Constitutional Critical Analysis of Socio-Juridical Changes in Current Cuba." *Revista Direito e Práxis* 8, no. 4: 3169–217.

Burnside Primary School. 2022. "Burnside Primary School BYOD Chromebooks 2022." Government of South Australia, Department for Education. Accessed May 23, 2023. https://www.burnsideps.sa.edu.au.

Burrows, Roger. 2012. "Living With the H-Index? Metric Assemblages in the Contemporary Academy." *Sociological Review* 60 (2): 355–72.

Bustos-Gallardo, Beatriz, Gavin Bridge, and Manuel Prieto. 2021. "Harvesting Lithium: Water, Brine and the Industrial Dynamics of Production in the Salar de Atacama." *Geoforum* 119:177–89.

Byler, Darren. 2022. *Terror Capitalism: Uyghur Dispossession and Masculinity in a Chinese City*. Durham, NC: Duke University Press.

Cachelin, Shala. 2022. "The US Drone Programme, Imperial Air Power and Pakistan's Federally Administered Tribal Areas." *Critical Studies on Terrorism* 15 (2): 441–62.

Caffentzis, George. 2013. *In Letters of Blood and Fire: Work, Machines, and the Crisis of Capitalism*. New York: PM Press.

Cancela, Ekaitz. 2023. *Utopías digitales: Imaginar el fin del capitalismo*. Barcelona: Verso.

Cancela, Ekaitz, and Aitor Jiménez. 2022. *Europe's Third Way to Technological Sovereignty: A Critique*. Brussels: Transform! Europe. https://www.transform-network.net.

Cancela, Ekaitz, and Stuart Medina. 2021. "Consultancy Capitalism Is Allowing Private Firms to Control Public Funds." *Jacobin*, November 8, 2021. https://jacobin.com.

Canning, Victoria, and Steve Tombs. 2021. *From Social Harm to Zemiology: A Critical Introduction*. London: Routledge.

Caplan, Jeremy. 2009. "Google and Microsoft: The Battle Over College E-Mail." *Time*, August 14, 2009. https://time.com/archive/.

Carrington, Damian. 2020. "World's Consumption of Materials Hits Record 100bn Tonnes a Year." *The Guardian*, January 22, 2020. https://www.theguardian.com.

Carroll, Rory. 2020. "Why Irish Data Centre Boom Is Complicating Climate Efforts." *The Guardian*, January 6, 2020. https://www.theguardian.com.

Castelos, Carla Noever. 2023. "Mining Out of the Crisis? The Role of the State in the Expansion of the Lithium Frontier in Extremadura, Spain." *Extractive Industries and Society* 15:101329.

Castillo, Elise. 2020. "A Neoliberal Grammar of Schooling? How a Progressive Charter School Moved Toward Market Values." *American Journal of Education* 126 (4): 519–47.

Castro-Gómez, Santiago. 2005. *La Hybris del Punto Cero: Ciencia, Raza e Ilustración en la Nueva Granada (1750–1816)*. Bogotá: Editorial Pontificia Universidad Javeriana.

Cellular Telecommunications and Internet Association. 2023. *Wireless Siting Reforms Drive Investment and Deployment Across the U.S.* Washington: CTIA. https://api.ctia.org.

Ceres, Pia. 2022. "Kids Are Back in Classrooms and Laptops Are Still Spying on Them." *Wired*, August 3, 2022. https://www.wired.com.

Chakrabortty, Aditya. 2022. "Why is Private Health Booming and the NHS in Crisis? Because that's What Ministers Want." *The Guardian*, October 13, 2022. https://www.theguardian.com.

Chamayou, Grégoire. 2021. *The Ungovernable Society: A Genealogy of Authoritarian Liberalism*. Hoboken, NJ: John Wiley & Sons.

Chan, Ngai Keung. 2019. "The Rating Game: The Discipline of Uber's User-Generated Ratings." *Surveillance & Society* 17 (1/2): 183–90.

Chander, Anupam, and Vivek Krishnamurthy. 2018. "The Myth of Platform Neutrality." *George Town Law Technology Review* 2 (2): 400–416.

Chun, Wendy Hui Kyong. 2021. *Discriminating Data: Correlation, Neighborhoods, and the New Politics of Recognition*. Cambridge, MA: MIT press.

Churchill, Ward, and Jim Vander Wall. 2002. *Agents of Repression: The FBI's Secret Wars against the Black Panther Party and the American Indian Movement*. London: South End.

City of Chicago. 2020. "Strategic Subject List—Historical Public Safety." Chicago Data Portal. Last Updated September 26, 2020. https://data.cityofchicago.org.

Clifford, Damian, Jake Goldenfein, Aitor Jiménez, and Megan Richardson. 2023. "A Right of Social Dialogue on Automated Decision-Making: From Workers' Right to Autonomous Right." *Technology and Regulation* 2023:1–9.

CNIL. 2022. "Cookies: the Council of State Confirms the 2020 Sanction Imposed by the CNIL Against Amazon." June 28, 2022. https://www.cnil.fr.

Colledge, Sue, James Conolly, Enrico Crema, and Stephen Shennan. 2019. "Neolithic Population Crash in Northwest Europe Associated with Agricultural Crisis." *Quaternary Research* 92, no. 3: 686–707.

Collington, Rosie. 2022. "Disrupting the Welfare State? Digitalisation and the Retrenchment of Public Sector Capacity." *New Political Economy* 27 (2): 312–28.

Collini, Stefan. 2017. *Speaking of Universities*. London: Verso Books.

Commonwealth of Australia. 2023. *Royal Commission into the Robodebt Scheme*. N.p.: Commonwealth of Australia. https://robodebt.royalcommission.gov.au.

CommScope. n.d. *Simplify and Evolve Your Mobile Network: A Guide for Optimizing Your RF Path*. Claremont: CommScope. https://www.commscope.com.

Connell, Raewyn. 2013. "The Neoliberal Cascade and Education: An Essay on the Market Agenda and Its Consequences." *Critical Studies in Education* 54 (2): 99–112.

Conger, Kate. 2021. "California's Gig Worker Law Is Unconstitutional, Judge Rules." *New York Times*, August 20, 2021. https://www.nytimes.com.

Couldry, Nick, and Ulises Ali Mejias. 2020. *The Costs of Connection: How Data is Colonizing Human Life and Appropriating It for Capitalism.* Stanford, CA: Stanford University Press.

Couldry, Nick, and Ulises Ali Mejias. 2023. "The Decolonial Turn in Data and Technology Research: What Is at Stake and Where Is It Heading?" *Information, Communication & Society* 26 (4): 786–802.

Coulthard, Glen Sean. 2014. *Red Skin, White Masks: Rejecting the Colonial Politics of Recognition.* Minneapolis: University of Minnesota Press.

Cowen, Deborah. 2020. "Following the Infrastructures of Empire: Notes on Cities, Settler Colonialism, and Method." *Urban Geography* 41 (4): 469–86.

Cram, W. Alec, Martin Wiener, Monideepa Tarafdar, and Alexander Benlian. 2022. "Examining the Impact of Algorithmic Control on Uber Drivers' Technostress." *Journal of Management Information Systems* 39 (2): 426–53.

Crawford, Kate. 2021. *Atlas of AI: Power, Politics, and the Planetary Costs of Artificial Intelligence.* New Haven, CT: Yale University Press.

Cristiano, Fabio. 2020. "Israel: Cyber Warfare and Security as National Trademarks of International Legitimacy." In *Routledge Companion to Global Cyber-Security Strategy,* edited by Scott N. Romaniuk and Mary Manjikian, 409–17. New York: Routledge.

Crook, Martin, and Damien Short. 2014. "Marx, Lemkin and the Genocide—Ecocide Nexus." *International Journal of Human Rights* 18, no. 3: 298–319.

Crook, Martin, Damien Short, and Nigel South. 2018. "Ecocide, Genocide, Capitalism and Colonialism: Consequences for Indigenous Peoples and Glocal Ecosystems Environments." *Theoretical Criminology* 22 (3): 298–317.

Cunneen, Christopher, and Juan Marcellus Tauri. 2016. *Indigenous Criminology.* Bristol, UK: Policy Press.

Cusicanqui, Silvia Rivera. 2012a. "Ch'ixinakax utxiwa: A Reflection on the Practices and Discourses of Decolonization." *South Atlantic Quarterly* 111 (1): 95–109.

Cusicanqui, Silvia Rivera. 2012b. *Violencia (Re) encubiertas en Bolivia.* La Paz, Bolivia: La Mirada Salvaje.

Daliri-Ngametua, Rafaan, and Ian Hardy. 2022. "The Devalued, Demoralized and Disappearing Teacher: The Nature and Effects of Datafication and Performativity in Schools." *Education Policy Analysis Archives* 30 (102): 1–24.

Dana, Tariq. 2020. "A Cruel Innovation: Israeli Experiments on Gaza's Great March of Return." *Sociology of Islam* 8 (2): 175–98.

Darling-Hammond, Linda 2007. "Race, Inequality and Educational Accountability: The Irony of 'No Child Left Behind.'" *Race Ethnicity and Education* 10 (3): 245–60.

Datatilsynet. 2022. "Datatilsynet nedlægger behandlingsforbud i Chromebook-sag." July 14, 2022. www.datatilsynet.dk.

Davidovic, Jovana, and Milton C. Regan Jr. 2023. *AI Weapons and Just Preparation for War.* N.p.: Babl AI and Stockdale Center for Ethical Leadership.

Davies, Harry, Bethan McKernan, and Dan Sabbagh. 2023. "'The Gospel': How Israel Uses AI to Select Bombing Targets in Gaza." *The Guardian,* December 1, 2023. https://www.theguardian.com.

Davies, Harry, Simon Goodley, Felicity Lawrence, Paul Lewis, and Lisa O'Carroll. 2022. "Uber Broke Laws, Duped Police and Secretly Lobbied Governments, Leak Reveals." *The Guardian*, July 11, 2022.
Davis, Angela Y. 2003. "Race and Criminalization: Black Americans and the Punishment Industry." In *Criminological Perspectives: Essential Readings*, edited by John Muncie, Eugene McLaughlin, and Gordon Hughes, 284–93. London: Sage.
Davis, Angela Y. 2011. *Are Prisons Obsolete?* New York: Seven Stories Press.
Davis, Muriam Haleh. 2022. *Markets of Civilization: Islam and Racial Capitalism in Algeria*. Durham, NC: Duke University Press.
de Cabo Martín, Carlos. 2022. *Pluralismo real (del Norte) y epistemología (del Sur) desde el constitucionalismo crítico*. Cizur Menor, Spain: Aranzadi/Civitas.
de la Torre Rangel, Jesús Antonio. 2006. *El derecho como arma de liberación en América Latina, sociología jurídica y uso alternativo del derecho*. San Luís Potosí, Mexico: Cenejus.
Defence Australia. "Information Warfare Division." YouTube video, 00:01:00, October 30, 2019. https://www.youtube.com/watch?v=mzq4XaDxfoA.
Del Mármol, Camila, and Ismael Vaccaro. 2020. "New Extractivism in European Rural Areas: How Twentieth First Century Mining Returned to Disturb the Rural Transition." *Geoforum* 116:42–49.
Delfanti, Alessandro. 2021a. "Machinic Dispossession and Augmented Despotism: Digital Work in an Amazon Warehouse." *New Media & Society* 23 (1): 39–55.
Delfanti, Alessandro. 2021b. *The Warehouse: Workers and Robots at Amazon*. London: Pluto Books.
Delio, Michelle. 2001. "Did IBM Help Nazis in WWII?" *Wired*, February 12, 2001. https://www.wired.com.
Department of Conservation. 2014. "Te Urewera Act 2014." July 27, 2014. https://www.legislation.govt.nz.
Diab, Ramon S. 2019. "Capital's Artificial Intellect Becoming Uber's Means of Autonomous Immaterial Production." *Historical Materialism* 27 (1): 125–54.
Digital Europe. 2022. *Digital Europe's Recommendations for the Critical Raw Materials Act*. Brussels: Digital Europe.
Djukanovic, Nina. 2022. "'Green Are Fields, Not Mines': The Case of Lithium Mining and Resistance in Serbia." PhD diss., University of Oxford.
Domínguez, Dani. 2021, "La relación entre el presidente extremeño Fernández Vara e Iberdrola." *La Marea*, November 8, 2021. https://www.lamarea.com.
Dubal, Veena B. 2017. "The Drive to Precarity: A Political History of Work, Regulation, & Labor Advocacy in San Francisco's Taxi & Uber Economies." *Berkeley Journal of Employment and Labor Law* 38 (1): 73–135.
Dubal, Veena B. 2020. "Digital Piecework." *Dissent* 67 (4): 37–44.
Dubal, Veena B. 2022. "Economic Security & the Regulation of Gig Work in California: From AB5 to Proposition 22." *European Labour Law Journal* 13 (1): 51–65.
Dube, Ryan. 2022. "The Place with the Most Lithium Is Blowing the Electric-Car Revolution." *Wall Street Journal*, August 10, 2022. https://www.wsj.com.

Duggan, Mark, Atul Gupta, Emilie Jackson, and Zachary S. Templeton. 2023. *The Impact of Privatization: Evidence from the Hospital Sector*. Cambridge, MA: National Bureau of Economic Research.

Dunlap, Alexander. 2021. "The Politics of Ecocide, Genocide and Megaprojects: Interrogating Natural Resource Extraction, Identity and the Normalization of Erasure." *Journal of Genocide Research* 23 (2): 212–35.

Dunlap, Alexander, and Mariana Riquito. 2023. "Social Warfare for Lithium Extraction? Open-Pit Lithium Mining, Counterinsurgency Tactics and Enforcing Green Extractivism in Northern Portugal." *Energy Research & Social Science* 95 (17): 102912.

Dyer-Witheford, Nick, and Svitlana Matviyenko. 2019. *Cyberwar and Revolution: Digital Subterfuge in Global Capitalism*. Minneapolis: University of Minnesota Press.

Dyer-Witheford, Nick, Atle Mikkola Kjøsen, and James Steinhoff. 2019. *Inhuman Power: Artificial Intelligence and the Future of Capitalism*. London: Pluto Press.

elDiarioAr. 2023. "Jujuy, las protestas y el negocio del litio: diez claves para entender qué pasa en la provincia." *ElDiarioAr*, June 26, 2023. https://www.eldiarioar.com.

Electronic Frontier Foundation (EFF). 2015. "Google Deceptively Tracks Students' Internet Browsing, EFF Says in FTC Complaint." EFF, December 1, 2015. https://www.eff.org.

Engstrom, David Freeman, Daniel E. Ho, Catherine M. Sharkey, and Mariano-Florentino Cuéllar. 2020. "Government by Algorithm: Artificial Intelligence in Federal Administrative Agencies." New York University School of Law, Public Law Research Paper No. 20-54.

Environmental Justice Atlas. 2021. "Lithium Mining in the Salar de Atacama, Chile." Environmental Justice Atlas, last updated April 25, 2023. https://www.ejatlas.org.

Enzesberger, Hans Magnus. 2020. *El corto verano de la anarquía*. Barcelona: Anagrama.

Ernst & Young. 2021 *The Twin Transition: A New Digital and Sustainability Framework for the Public Sector*. N.p.: Microsoft Corporation. https://wwps.microsoft.com.

Escobar, Arturo. 2011. "Ecología política de la globalidad y la diferencia." In *La naturaleza colonizada: Ecología política y minería en América Latina*, edited by Héctor Alimonda, 61–92. Buenos Aires: Clacso.

Eubanks, Virginia. 2018. *Automating Inequality: How High-Tech Tools Profile, Police, and Punish the Poor*. New York: St. Martin's Press.

Eurojust. 2021. *Report on Eurojust's Casework on Environmental Crime*. N.p.: European Union Agency for Criminal Justice Cooperation.

European Data Protection Board. 2023. "Decision of the Data Protection Commission Made Pursuant to Section 111 of the Data Protection Act, 2018 and Articles 60 and 65 of the General Data Protection Regulation." European Data Protection Board, May 22, 2023. https://edpb.europa.eu.

European Commission. 2017. "Antitrust: Commission fines Google €2.42 Billion for Abusing Dominance as Search Engine by Giving Illegal Advantage to Own Comparison Shopping Service—Factsheet." June 27, 2017. https://ec.europa.eu.

European Commission. 2018. "Antitrust: Commission Fines Google €4.34 Billion for Illegal Practices Regarding Android Mobile Devices to Strengthen Dominance of Google's Search Engine." Press release, July 18, 2018. https://ec.europa.eu.

European Commission. 2019a. "Antitrust: Commission Fines Google €1.49 Billion for Abusive Practices in Online Advertising." Press release, March 20, 2019.

European Commission. 2019b. "Communication from the Commission: The European Green Deal." EUR-Lex: Access to European Union Law, December 11, 2019. https://eur-lex.europa.eu.

European Commission. 2020. "Communication from the Commission to the European Parliament, the Council, the European Economic and Social Committee and the Committee of the Region—Critical Raw Materials Resilience: Charting a Path towards Greater Security and Sustainability." EUR-Lex: Access to European Union Law, September 3, 2020. https://eur-lex.europa.eu.

European Commission. 2021. "Proposal for a Directive of the European Parliament and of the Council on the Protection of the Environment through Criminal Law and Replacing Directive 2008/99/EC." EUR-Lex: Access to European Union Law, December 15, 2021. https://commission.europa.eu.

European Commission. 2023a. "Communication from the Commission to the European Parliament, the Council, the European Economic and Social Committee and the Committee of the Regions—A Secure and Sustainable Supply of Critical Raw Materials in Support of the Twin Transition." EUR-Lex: Access to European Union Law, March 16, 2023. https://eur-lex.europa.eu.

European Commission. 2023b. "Proposal for a regulation of the European Parliament and of the Council Establishing a Framework for Ensuring a Secure and Sustainable Supply of Critical Raw Materials and Amending Regulations (EU) 168/2013, (EU) 2018/858, 2018/1724 and (EU) 2019/1020." EUR-Lex: Access to European Union Law, March 16, 2023. https://eur-lex.europa.eu.

European Commission. 2024a. "Countering Irregular Migration: Better EU Border Management." European Commission, April 16, 2024. https://home-affairs.ec.europa.eu.

European Commission. 2024b. "Integrated Border Management Fund—Border Management and Visa Instrument (2021–27)." June 6, 2024. https://home-affairs.ec.europa.eu.

European Council. 2024. "Migration Flows on the Western Routes." January 11, 2024. https://www.consilium.europa.eu.

European Court of Justice. 2017. "Press Release No 30/17, Judgments in Cases C-157/15." Luxembourg, March 14, 2017. https://curia.europa.eu.

European Parliament. 2023a. "Fit For 55: Zero CO_2 Emissions for New Cars and Vans in 2035." Press release February 14, 2023. https://www.europarl.europa.eu.

European Parliament. 2023b. "Revision of Directive 2008/99/EC Protection of the Environment through Criminal Law." Think Tank: European Parliament. December 13, 2023. https://www.europarl.europa.eu.

European Round Table for Industry. 2022. *ERT Expert Paper—Action Plan for a Digitally Enabled Green Transition*. Prepared by the Task Force on the Digitally Enabled Green Transition. Brussels: European Round Table for Industry. https://ert.eu.
Europol. 2022. "Enviromental Crime." Accessed May 5, 2023. https://www.europol.europa.eu.
Evangelista, Rafael, and Fernanda Bruno. 2019. "Whatsapp and Political Instability in Brazil: Targeted Messages and Political Radicalisation." *Internet Policy Review* 8 (4): 1–23.
Extinction Rebellion. 2019. *This Is Not a Drill: An Extinction Rebellion Handbook*. Edited by Clare Farrell, Alison Green, Sam Knights, and William Skeaping. London: Penguin.
Fajardo del Castillo, Teresa. 2015. *EU Environmental Law and Environmental Crime: An Introduction*. Granada: University of Granada.
Farwell, James P., and Rafal Rohozinski. 2011. "Stuxnet and the Future of Cyber War." *Survival* 53 (1): 23–40.
Federal Trade Commission. 2019a. "FTC Imposes $5 Billion Penalty and Sweeping New Privacy Restrictions on Facebook." July 23, 2019.
Federal Trade Commission. 2019b. "Google and YouTube Will Pay Record $170 Million for Alleged Violations of Children's Privacy Law." September 4, 2019.
Federal Trade Commission. 2021. "FTC to Ramp Up Law Enforcement Against Illegal Repair Restrictions." July 21, 2021. www.ftc.gov.
Federici, Silvia. 2004. *Caliban and the Witch*. New York: Autonomedia.
Federici, Silvia. 2006. "Precarious Labour: A Feminist Viewpoint." Lecture given at Bluestockings Radical Bookstore, New York City, October 28, 2006.
Feldman, Zeena, and Marisol Sandoval. 2018. "Metric Power and the Academic Self: Neoliberalism, Knowledge and Resistance in the British University." *TripleC: Communication, Capitalism & Critique* 16 (1): 214–33.
Fernández, Gonzalo, Erika González, and Pedro Ramiro. 2022. *El Boom minero: Patrones e impactos de la expansión de la industria extractiva en España*. Bilbao, Spain: OMAL and Amigos de la Tierra.
Fernández, Gonzalo, Erika González, Juan Hernández, and Pedro Ramiro. 2022. "Megaproyectos: Claves de análisis y resistencia en el capitalismo verde y digital." Bilbao, Spain: OMAL. https://omal.info.
Fitzmaurice, Andrew. 2007. "The Genealogy of *Terra Nullius*." *Australian Historical Studies* 38 (129): 1–15.
Food and Agriculture Organization (FAO). 2022. "The Status of Fishery Resources." In *The State of World Fisheries and Aquaculture 2022: Towards Blue Transformation*. Rome: FAO.
Foucault, Michel. 2008. *The Birth of Biopolitics: Lectures at the Collège de France, 1978–1979*. Edited by Michel Senellart, Arnold I. Davidson Alessandro Fontana, and Francois Ewald. Tranlated by Graham Burchell. Berlin: Springer.
French, Duncan, and Louis J. Kotzé. 2019. "'Towards a Global Pact for the Environment': International Environmental Law's Factual, Technical and (Unmentionable)

Normative Gaps." *Review of European, Comparative & International Environmental Law* 28 (1): 25–32.

Friedman, Milton. 1962. *Capitalism and Freedom*. Chicago: Chicago University Press.

Friedman, Milton. 1970. "A Friedman Doctrine: The Social Responsibility of Business is to Increase its Profits." *New York Times*, September 13, 1970, 32–33.

Friends of the Earth Europe and Rachel Tansey. 2023. *Mining the Depths of Influence: How Industry is Forging the EU Critical Raw Materials Act*. N.p.: Friends of the Earth Europe. https://friendsoftheearth.eu.

Fuchs, Christian. 2018. "Universal Alienation, Formal and Real Subsumption of Society under Capital, Ongoing Primitive Accumulation by Dispossession: Reflections on the Marx@ 200-contributions by David Harvey and Michael Hardt/Toni Negri." *tripleC: Communication, Capitalism & Critique* 16 (2): 454–67.

Gago, Verónica. 2017. *Neoliberalism from Below: Popular Pragmatics and Baroque Economies*. Durham, NC: Duke University Press.

Gajardo, Gonzalo, and Stella Redón. 2019. "Andean Hypersaline Lakes in the Atacama Desert, Northern Chile: Between Lithium Exploitation and Unique Biodiversity Conservation." *Conservation Science and Practice* 1 (9): e94.

Galaz-Mandakovic, Damir, Víctor Tapia Araya, and Francisco Rivera. 2023. "New Historical Archives of Extractivism in the Atacama Desert: Contamination and Mortality during the Guggenheim Period in Chuquicamata, Chile, 1915–1923." *Extractive Industries and Society* 13:101202.

Gallagher, Jill C., and Nicole T. Carter. 2023. *Protection of Undersea Telecommunication Cables: Issues for Congress*. N.p.: Congressional Research Service. https://crsreports.congress.gov.

Gallagher, Zoe. 2023. "Will the Real Eco-Terrorists Please Stand Up?" *Hastings Environmental Law Journal* 29 (1): 27–46.

García, Sergio García, Ignacio Mendiola, Débora Ávila, Laurent Bonelli, José Ángel Brandariz, Cristina Fernández Bessa, and Manuel Maroto Calatayud. 2021. *Metropolice: Seguridad y policía en la ciudad neoliberal*. Madrid: Traficantes de Sueños.

Gargarella, Roberto. 2013. *Latin American Constitutionalism, 1810–2010: The Engine Room of the Constitution*. Oxford: Oxford University Press.

Gault, Matthew. 2021. "Amazon Won't Let Us Leave." *Vice*, December 18, 2021. https://www.vice.com.

Gebrial, Dalia. 2022. "Racial Platform Capitalism: Empire, Migration and the Making of Uber in London." *Environment and Planning A: Economy and Space* 0 (1): 1–25.

Gebru, Timnit. 2020. "Race and Gender." In *The Oxford Handbook of Ethics of AI*, edited by Markus D. Dubber, Frank Pasquale and Sunit Das, 251–69. Oxford: Oxford University Press, 2020.

Geddis, Andrew, and Jacinta Ruru. 2019. "Places as Persons: Creating a New Framework for Māori-Crown Relations." In *The Frontiers of Public Law*, edited by Jason Varuhas, 298–316. Oxford: Hart Publishing.

Geiger, Gabriel. 2021. "How a Discriminatory Algorithm Wrongly Accused Thousands of Families of Fraud." *Vice*, March 2, 2021. https://www.vice.com.
Gig Economy Project and Nuria Soto. 2021. "Gig Economy Project—'We Are What the Companies Don't Want to Exist': Interview with Nuria Soto of 'RidersXDerechos.'" *Brave New Europe*, November 2, 2021. https://braveneweurope.com.
Gilliom, John. 2001. *Overseers of the Poor: Surveillance, Resistance, and the Limits of Privacy*. Chicago: University of Chicago Press.
Gilmore, Ruth Wilson. 2002. "Fatal Couplings of Power and Difference: Notes on Racism and Geography." *Professional Geographer* 54, no. 1: 15–24.
Gilmore, Ruth Wilson. 2007. *Golden Gulag: Prisons, Surplus, Crisis, and Opposition in Globalizing California*. Berkeley: University of California Press.
Gilmore, Ruth Wilson. 2022. *Abolition Geography: Essays towards Liberation*. London: Verso Books.
Gilroy, Paul. 2013. *There Ain't No Black in the Union Jack*. London: Routledge.
Global Foodprint Network. 2023. "Open Data Platform—Global Footprint Network." Accessed May 26, 2023. https://data.footprintnetwork.org.
Global Forest Watch. 2023. "Dashboards." Accessed March 30, 2023. https://www.globalforestwatch.org.
Goikoetxea, Jule. 2024. "Idealism and Biologism in Social Reproduction Theory: A Materialist Critique." *Capital & Class*, preprint, submitted February 6, 2024.
Goldenfein, Jake, and Daniel Griffin. 2022. "Google Scholar—Platforming the Scholarly Wconomy." *Internet Policy Review* 11 (3).
Goldman Sachs. 2023. "Electric Vehicles Are Forecast to Be Half of Global Car Sales by 2035." February 10, 2023. https://www.goldmansachs.com.
Gómez, Hector (@Hectorgomezh). 2023. "Junto a @GFVara, he mantenido un encuentro muy positivo con empresas en #Extremadura. Suponen futuro, transformación y sostenibilidad en esta tierra." Twitter, May 15, 2023, 10:42 p.m. https://twitter.com.
Gonzaga, Diego. 2022. "4 Years of Amazon Destruction." Greenpeace, December 2, 2022. https://www.greenpeace.org.
González, Roberto J. 2022. *War Virtually: The Quest to Automate Conflict, Militarize Data, and Predict the Future*. Berkeley: University of California Press.
Goodfriend, Sophia. 2023. "Algorithmic State Violence: Automated Surveillance and Palestinian Dispossession in Hebron's Old City." *International Journal of Middle East Studies* 55 (3): 461–78.
Google Educator Groups Spain. 2022. "III Trobada Col·la~Clic (Antic G Suite Catalunya)." GEG Spain Comunidad Profesional [GEG Spain Professional Community]. October 12, 2022. https://www.gedu.es.
Google Employees. 2018. "We Work for Google. Our Employer Shouldn't Be in the Business of War." *The Guardian*, April 5, 2018. www.theguardian.com.
Google. 2020. "Fundamentals Training." August 28, 2020. https://skillshop.exceedlms.com.
Google. 2021. "A Peek at What's Next for Google Classroom." *Google: The Keyword* (blog). February 17, 2021. https://blog.google.

Google. 2022. "24/7 Carbon-Free Energy: Powering Up New Clean Energy Projects across the Globe." Google Cloud. April 21, 2022. https://cloud.google.com.

Google. n.d. "Engage, Inspire, and Learn from a Passionate Community of Educators." Accessed February 15, 2023. https://edu.google.com.

Gorwa, Robert. 2019. "What is Platform Governance?" *Information, Communication & Society* 22 (6): 854–71.

Gorz, André. 1994. *Capitalism, Socialism, Ecology*. London: Verso Books.

Goyes, David Rodríguez. 2019. *Southern Green Criminology: A Science to End Ecological Discrimination*. Bradford: Emerald Publishing.

Grandinetti, Justin. 2022. "'From the Classroom to the Cloud': Zoom and the Platformization of Higher Education." *First Monday* 27 (2).

Granick, Jennifer Stisa. 2017. *American Spies: Modern Surveillance, Why You Should Care, and What to Do About It*. Cambridge: Cambridge University Press.

Green, Ben. 2022. "The Flaws of Policies Requiring Human Oversight of Government Algorithms." *Computer Law & Security Review* 45:105681.

Greenwald, Glenn, and Ewen MacAskill. 2013. "NSA Prism Program Taps into User Data of Apple, Google and Others." *The Guardian*, June 7, 2013.

Grewal, David Singh, and Jedediah S. Purdy. 2014. "Introduction: Law and Neoliberalism." *Law and Contemporary Problems* 77 (4): 1–23.

Griffin, Paul, and Richard Heede. 2017. *The Carbon Majors Database: CDP Carbon Majors Report 2017*. London: CDP Worldwide.

Grosfoguel, Ramon. 2016. "What Is Racism?" *Journal of World-Systems Research* 22 (1): 9–15.

Grothoff, Christian, and Jens M. Porup. 2016. "The NSA's SKYNET Program May Be Killing Thousands of Innocent People." *Ars Technica*, February 16, 2016. https://arstechnica.com.

Gudynas, Eduardo. 2021. *Extractivisms: Politics, Economy and Ecology*. Black Point, CA: Fernwood.

Guterres, António. 2023. "Secretary-General's Video Message to the WMO 'State of the Global Climate 2023' Report launch." United Nations. November 30, 2023. https://www.un.org.

Gutiérrez, Jorge S., Johnnie N. Moore, J. Patrick Donnelly, Cristina Dorador, Juan G. Navedo, and Nathan R. Senner. 2022. "Climate Change and Lithium Mining Influence Flamingo Abundance in the Lithium Triangle." *Proceedings of the Royal Society B* 289 (1970): 20212388.

Hacking, Ian. 1990. *The Taming of Chance*. Cambridge: Cambridge University Press.

Haidt, Jonathan. 2021. "The Dangerous Experiment on Teen Girls." *The Atlantic*, November 21, 2021.

Haines, Fiona, and Christine Parker. 2017. "Moving towards Ecological Regulation: The Role of Criminalisation." In *Criminology and the Anthropocene*, edited by Cameron Holley and Clifford Shearing, 81–108. London: Routledge.

Hajjat, Abdellali, and Marwan Mohammed. 2023. *Islamophobia in France: The Construction of the "Muslim Problem."* Athens: University of Georgia Press.

Hamilton, Laura T., Heather Daniels, Christian Michael Smith, and Charlie Eaton. 2022. "The Private Side of Public Universities: Third-Party Providers and Platform Capitalism." UC Berkeley Center for Studies in Higher Education, Research & Occasional Paper Series.

Hampton, Lelia Marie. 2021. "Black Feminist Musings on Algorithmic Oppression." *arXiv*, preprint, submitted January 25, 2021.

Hancox, Dan. 2020. "The 2010 Student Protests Were Vilified—but Their Warnings of Austerity Britain Were Proved Right." *The Guardian*, November 12, 2020. https://www.theguardian.com.

Haraway, Donna J. 2016. *Manifestly Haraway*. Minneapolis: University of Minessota Press.

Harper, Krista. 2005. "'Wild Capitalism' and 'Ecocolonialism': A Tale of Two Rivers." *American Anthropologist* 107 (2): 221–33.

Harvey, David. 2007. *A Brief History of Neoliberalism*. Oxford: Oxford University Press.

Haskins, Caroline. 2024. "Google Fires 28 Workers for Protesting Cloud Deal with Israel." *Wired*, April 18, 2024. www.wired.com.

Hassan, Samer, and Primavera De Filippi. 2017. "The Expansion of Algorithmic Governance: From Code Is Law to Law Is Code." Special issue, *Field Actions Science Reports* 17: 88–90.

Haugen, Frances. 2021a. "Live: Facebook Whistleblower Testifies in European Parliament." Frances Haugen testifies at European Parliament. YouTube Video, 02:56:00, November 8, 2021. https://www.youtube.com/watch?v=Qlae1LnYngM.

Haugen, Frances. 2021b. "Sub-Committee on Consumer Protection, Product Safety, and Data Security. Statement of Frances Haugen." October 4, 2021. https://www.commerce.senate.gov.

Heal, Alexandra, Andrew Wasley, Sam Cutler, and André Campos. 2019. "Revealed: Fires Three Times More Common in Amazon Beef Farming Zones." *The Guardian*, December 10, 2019.

Heikkilä, Melissa. 2022. "Dutch Scandal Serves as a Warning for Europe over Risks of Using Algorithms." *Politico*, March 29, 2022. https://www.politico.eu.

Henley, Jon. 2021. "Dutch Government Resigns Over Child Benefits Scandal." *The Guardian*, January 15, 2021. https://www.theguardian.com.

Hern, Alex. 2014. "Google Faces Lawsuit Over Email Scanning and Student Data." *The Guardian*. March 19, 2014. https://www.theguardian.com.

Hernandez, Daniela Soto, and Peter Newell. 2022. "Oro Blanco: Assembling Extractivism in the Lithium Triangle." *Journal of Peasant Studies* 49 (5): 945–68.

Higgins, Rawinia. 2018. "Ko te mana tuatoru, ko te mana motuhake." In *Indigenous Peoples and the State: International Perspectives on the Treaty of Waitangi*, edited by Mark Hickford and Carwyn Jones, 129–39. London: Routledge.

Hildebrandt, Mireille. 2018. "Algorithmic Regulation and the Rule of Law." *Philosophical Transactions of the Royal Society A: Mathematical, Physical and Engineering Sciences* 376 (2128): 20170355.

Hillyard, Paddy, and Steve Tombs. 2007. "From 'Crime' to Social Harm?" *Crime, Law and Social Change* 48 (1–2): 9–25.

Holon IQ. 2021. "$16.1B of Global EdTech Venture Capital in 2020." Education Intelligence Unit, January 5, 2021.

Horne, Gerald. 2014. *The Counter-Revolution of 1776. Slave Resistance and the Origins of the United States of America.* New York: New York University Press.

Horwitz, Jeff. 2021. "Facebook Says Its Rules Apply to All. Company Documents Reveal a Secret Elite That's Exempt." *Wall Street Journal*, September 13, 2021.

Horwitz, Jeff, Deepa Seetharaman, and Georgia Wells. 2021. "Facebook Knows Instagram is Toxic for Teen Girls, Company Documents Show." *Wall Street Journal*, September 14, 2021.

House of Lords. 2021. *Digital Regulation: Joined-Up and Accountable.* Communications and Digital Committee. 3rd Report of Session 2021–22. https://committees.parliament.uk.

Humber, Lee Anderson. 2016. "The Impact of Neoliberal Market Relations of the Production of Care on the Quantity and Quality of Support for People with Learning Disabilities." *Critical and Radical Social Work* 4 (2): 149–67.

Hummel, Tassilo. 2023. "French Court Orders Uber to Pay Some $18 Mln to Drivers, Company to Appeal." *Reuters*, January 21, 2023. https://www.reuters.com.

Hund, Kirsten, Daniele La Porta, Thao P. Fabregas, Tim Laing, and John Drexhage. 2020. "Minerals for Climate Action: The Mineral Intensity of the Clean Energy Transition." Washington, DC: International Bank for Reconstruction and Development and The World Bank, 2020.

Hursh, David. 2001. "Social Studies within the Neo-Liberal State: The Commodification of Knowledge and the End of Imagination." *Theory & Research in Social Education* 29 (2): 349–56.

Hursh, David, and Camille Anne Martina. 2003. "Neoliberalism and Schooling in the US: How State and Federal Government Education Policies Perpetuate Inequality." *Journal for Critical Education Policy Studies* 1 (2): 1–13.

Infinity Lithium. n.d. "Home." Accesed June 24, 2024. https://www.infinitylithium.com.

Intergovernmental Panel on Climate Change. 2021. *Climate Change 2021: The Physical Science Basis; Working Group I Contribution to the Sixth Assessment Report of the Intergovernmental Panel on Climate Change.* Cambridge: Cambridge University Press.

International Energy Agency. 2021. *The Role of Critical Minerals in Clean Energy Transitions.* Paris: IEA Publications. https://www.iea.org.

International Energy Agency. 2023. "Final List of Critical Minerals 2022." Last updated February 3, 2023. https://www.iea.org.

International Monetary Fund. 2021. *International Monetary Fund Annual Report 2021.* N.p.: International Monetary Fund.

International Organization for Migration. 2022. "More than 5,000 Deaths Recorded on European Migration Routes since 2021: IOM." UN Migration. October 25, 2022. https://www.iom.int.

Inuarak, Naymen. 2022. "Steel Giant Arcelormittal Must Oppose Canadian Mine Expansion to Protect Inuit Way of Life." *Greenpeace*, January 11, 2022. https://www.greenpeace.org.

Ivancheva, Mariya, and Brian Garvey. 2022. "Putting the University to Work: The Subsumption of Academic Labour in UK's Shift to Digital Higher Education." *New Technology, Work and Employment* 37 (3): 381–97.

Jackson, Moana. 2018. "Colonization as Myth-Making: A Case Study in Aotearoa." In *Being Indigenous*, edited by Neyooxet Greymorning, 89–101. London: Routledge.

Jamil, Rabih. 2020. "Uber and the Making of an Algopticon-Insights from the Daily Life of Montreal Drivers." *Capital & Class* 44 (2): 241–60.

Jassy, Andy. 2022. "2021 Letter to Shareholders." Amazon. April 14, 2022. https://www.aboutamazon.com.

Jayasuriya, Kanishka. 2021. "COVID-19, Markets and the Crisis of the Higher Education Regulatory State: The Case of Australia." *Globalizations* 18 (4): 584–99.

Jefferson, Brian. 2020. *Digitize and Punish: Racial Criminalization in the Digital Age*. Minneapolis: University of Minnesota Press.

Jerez, Bárbara, Ingrid Garcés, and Robinson Torres. "Lithium Extractivism and Water Injustices in the Salar de Atacama, Chile: The Colonial Shadow of Green Electromobility." *Political Geography* 87: 102382.

Jewish Diaspora in Tech. 2022. "Palestinian, Jewish, Muslim, and Arab Google Employees Speak Out." Accessed April 18, 2023. https://jewishdiasporatech.org.

Jiménez, Aitor. 2020. "The Silicon Doctrine." *TripleC: Communication, Capitalism & Critique* 18 (1): 322–36.

Jiménez, Aitor. 2022. "Law, Code and Exploitation: How Corporations Regulate the Working Conditions of the Digital Proletariat." *Critical Sociology* 48 (2): 361–73.

Jiménez, Aitor, and Ekaitz Cancela. 2023. "¿Es posible gobernar a las plataformas digitales?: Análisis crítico de la Ley Europea de Servicios Digitales." *Teknokultura: Revista de Cultura Digital y Movimientos Sociales* 20 (1): 91–99.

Jiménez, Aitor, and Ainhoa Nadia Douhaibi. 2023. "The Islamophobic Consensus. Datafying Racism in Catalonia." In *Money, Power, and AI: Automated Banks and Automated States*, edited by Zofia Bednarz and Monika Zalnieriute, 152–70. Cambridge: Cambridge University Press.

Jiménez, Aitor, and James C. Oleson. 2022. "The Crimes of Digital Capitalism." *Mitchell Hamline Law Review* 48 (4): 971–1018.

Joint Research Centre. 2022. "The Twin Green & Digital Transition: How Sustainable Digital Technologies Could Enable a Carbon-Neutral EU by 2050." EU Science Hub, June 29, 2022. https://joint-research-centre.ec.europa.eu.

Joint Research Centre. 2023. "Severe Drought: Western Mediterranean Faces Low River Flows and Crop Yields Earlier Than Ever." EU Science Hub, June 13, 2023. https://joint-research-centre.ec.europa.eu.

Jougleux, Philippe. 2022. *Facebook and the (EU) Law: How the Social Network Reshaped the Legal Framework*. Berlin: Springer.

Junta de Extremadura. 2022. "Anuncio de 27 de diciembre de 2022. Consejería para la Transición Ecológica y Sostenibilidad." DOE Número 249, December 30, 2022. https://doe.juntaex.es.

Kaldani, Tamar, and Zeev Prokopets. 2022. *Pegasus Spyware and Its Impacts on Human Rights*. N.p.: Council of Europe, Information Society Department. https://rm.coe.int.

Kalpagam, Umamaheswaran. 2014. *Rule by Numbers: Governmentality in Colonial India*. Lanham, MD: Lexington Books.

Kalsnes, Bente, and Karoline Andrea Ihlebæk. 2021. "Hiding Hate Speech: Political Moderation on Facebook." *Media, Culture & Society* 43 (2): 326–42.

Kapustin, Max, Jens Ludwig, Marc Punkay, Kimberley Smith, Lauren Speigel, and David Welgus. 2017. *Gun Violence in Chicago, 2016*. Chicago: University of Chicago Crime Lab.

Karp, Paul. 2023. "Scott Morrison Accuses Labor of Campaign of 'Political Lynching' Against Him on Robodebt." *The Guardian*, July 31, 2023. www.theguardian.com.

Kassam, Ashifa. 2022. "'A Bloodbath': Refugees Reel from Deadly Melilla Mass Crossing." *The Guardian*, June 30, 2022. www.theguardian.com.

Kassem, Sarrah. 2022. "(Re)shaping Amazon Labour Struggles on Both Sides of the Atlantic: The Power Dynamics in Germany and the US Amidst the Pandemic." *Transfer: European Review of Labour and Research* 28 (4): 441–56.

Kastrenakes, Jacob. 2014. "Google Offers Schools Unlimited Drive Storage for Students and Teachers." *The Verge*, September 30, 2014. www.theverge.com.

Katwala, Amit. 2022. "The World Can't Wean Itself Off Chinese Lithium." *Wired*, June 30, 2022. www.wired.co.uk.

Kerssens, Niels, and José van Dijck. 2021. "The Platformization of Primary Education in the Netherlands." *Learning, Media and Technology* 46 (3): 250–63.

Khan, Azmat. 2021. "Hidden Pentagon Records Reveal Patterns of Failure in Deadly Airstrikes." *New York Times*, December 18, 2021. https://www.nytimes.com.

Kilgore, James. 2022. *Understanding E-Carceration: Electronic Monitoring, the Surveillance State, and the Future of Mass Incarceration*. New York: New Press.

King-Jones, Abi, and Erroll Wright. 2011. *Operation 8: Deep in the Forest*. Documentary. Wellington, New Zealand: Cutcutcut Films.

Kirchgaessner, Stephanie, and Sam Jones. 2020. "Phone of Top Catalan Politician 'Targeted by Government-Grade Spyware." *The Guardian*, July 13, 2020. www.theguardian.com.

Kirchman, David L. 2021. *Dead Zones: The Loss of Oxygen from Rivers, Lakes, Seas, and the Ocean*. Oxford: Oxford University Press.

Klein, Naomi. 2020. "Screen New Deal: Unver Cover of Mass Death, Andrew Cuomo Calls in the Billionaires to Build a High-Tech Dystopia." *The Intercept*, May 8, 2020. https://theintercept.com.

Klossa, Guillaume. 2019. *Towards European Media Sovereignty: An Industrial Media Strategy to Leverage Data, Algorithms, and Artificial Intelligence*. Brussels: European Commission.

Knight, Catherine. 2009. "The Paradox of Discourse Concerning Deforestation in New Zealand: A Historical Survey." *Environment and History* 15 (3): 323–42.
Knox, Robert. 2023. "International Law, Race, and Capitalism: A Marxist Perspective." *AJIL Unbound* 117:55–60.
Kramer, Ronald C. 1984. "Is Corporate Crime Serious Crime? Criminal Justice and Corporate Crime Control." *Journal of Contemporary Criminal Justice* 2 (3): 7–10.
Krutka, Daniel G., Ryan M. Smits, and Troy A. Willhelm. 2021. "Don't Be Evil: Should We Use Google in Schools?" *TechTrends* 65:421–31.
Kukutai, Tahu, and John Taylor. 2016. *Indigenous Data Sovereignty: Toward an Agenda*. Canberra, Australia: Australian National University Press.
Kumar, Priya C., Jessica Vitak, Marshini Chetty, and Tamara L. Clegg. 2019. "The Platformization of the Classroom: Teachers as Surveillant Consumers." *Surveillance & Society* 17 (1/2): 145–52.
Kundnani, Arun. 2014. *The Muslims Are Coming!: Islamophobia, Extremism, and The Domestic War on Terror*. London: Verso Books.
Kundnani, Arun. 2021. "The Racial Constitution of Neoliberalism." *Race & Class* 63 (1): 51–69.
Kwet, Michael. 2019. "Digital Colonialism: US Empire and the New Imperialism in the Global South." *Race & Class* 6 (4): 3–26.
Kwet, Michael. 2022. "The Digital Tech Deal: A Socialist Framework for the Twenty-First Century." *Race & Class* 63 (3): 63–84.
Laird, Elizabeth, Hugh Grant-Chapman, Cody Venzke, and Hannah Quay-de la Vallee. 2022. *Hidden Harms: The Misleading Promise of Monitoring Students Online*. Washington, DC: Center for Democracy and Technology. https://cdt.org.
Lasslett, Kristian. 2014. *State Crime on the Margins of Empire: Rio Tinto, the War on Bougainville and Resistance to Mining*. London: Pluto Press.
Lata, Lutfun Nahar, Jasmine Burdon, and Tim Reddel. 2023. "New Tech, Old Exploitation: Gig Economy, Algorithmic Control and Migrant Labour." *Sociology Compass* 17 (1): e13028.
Lechanteaux, Julie. 2021. "Role of the Big Four in EU Policymaking." Parliamentary question—E-001833/2021. Accessed May 26, 2023. www.europarl.europa.eu.
Lerner, Steve. 2012. *Sacrifice Zones: The Front Lines of Toxic Chemical Exposure in the United States*. Cambridge, MA: MIT Press.
Lessig, Lawrence. 2006. *CODE Version 2.0*. New York: Basic Books.
Leufer, Daniel, Caterina Rodelli, and Fanny Hidvegi. 2023. "Human Rights Protections ... with Exceptions: What's (Not) In the EU's AI Act Deal." *Access Now*, December 14, 2023.
Li, Pengfei, Jianyi Yang, Mohammad A. Islam, and Shaolei Ren. 2023. "'Making Ai Less Thirsty': Uncovering and Addressing the Secret Water Footprint of Ai Models." *arXiv* preprint, submitted April 6, 2023.
Libicki, Martin C. 2009. *Cyberdeterrence and Cyberwar*. Santa Monica: RAND Corporation.

Liedke, Jacob, and Luxuan Wang. 2023. "News Platform Fact Sheet." *Pew Research Center*, November 15, 2023. www.pewresearch.org.
Lilli, Eugenio. 2020. "President Obama and US Cyber Security Policy." *Journal of Cyber Policy* 5 (2): 265–84.
Linera, Álvaro García. 2022. "Conversatorio con Álvaro García Linera." Celag Data, January 22, 2020. www.celag.org.
Linera, Álvaro García, and Pablo Stefanoni. 2008. *La potencia plebeya: Acción colectiva e identidades indígenas, obreras y populares en Bolivia*. Buenos Aires: Prometeo Libros Editorial.
Lipton, Beryl. 2020. "Predictive Algorithms, Big Data Analytics, and 'Smart Technologies' Deployed by Governments in the U.S." MuckRock. Accesed June 24, 2024. www.muckrock.com.
Lithium Iberia. n.d. a. "The Project." Accesed May 13, 2023. https://lithiumiberia.com.
Lithium Iberia. n.d. b. "Green Mining." Accesed May 15, 2023. https://lithiumiberia.com.
Lithium Iberia. 2022. "Lithium Iberia patrocina por tercer año consecutivo al Club de Baloncesto Sagrado Corazón de Cáceres que pasará a denominarse Lithium Iberia-Sagrado Corazón." July 6, 2022. https://lithiumiberia.com.
Liu, Wenjuan, and Datu B. Agusdinata. 2020. "Interdependencies of Lithium Mining and Communities Sustainability in Salar de Atacama, Chile." *Journal of Cleaner Production* 260:120838.
LobbyFacts. 2024. "European Round Table for Industry." April 5, 2024. https://www.lobbyfacts.eu.
Loewenstein, Antony. 2023. *The Palestine Laboratory: How Israel Exports the Technology of Occupation Around the World*. London: Verso Books.
Lorenz, Chris. 2012. "If You're so Smart, Why Are You Under Surveillance? Universities, Neoliberalism, and New Public Management." *Critical Inquiry* 38 (3): 599–629.
Lorey, Isabell. 2015. *State of Insecurity: Government of the Precarious*. London: Verso.
Lupton, Deborah, and Ben Williamson. 2017. "The Datafied Child: The Dataveillance of Children and Implications for Their Rights." *New Media & Society* 19 (5): 780–94.
Lyon, David. 2015. *Surveillance after Snowden*. Hoboken: John Wiley & Sons.
Maki, Krystle. 2011. "Neoliberal Deviants and Surveillance: Welfare Recipients under the Watchful Eye of Ontario Works." *Surveillance & Society* 9 (1/2): 47–63.
Malm, Andreas. 2016. *Fossil Capital: The Rise of Steam Power and the Roots of Global Warming*. London: Verso Books.
Mankelow, Natalie. 2014. "Police Apologise to Tuhoe Over Raids." *RNZ*, August 13, 2014. https://www.rnz.co.nz.
Mann, Monique. 2020. "Technological Politics of Automated Welfare Surveillance: Social (and Data) Justice through Critical Qualitative Inquiry." *Global Perspectives* 1 (1).
Marinetti, Filippo Tommaso. 2008. "The Manifesto of Futurism." *The Guardian*, November 13, 2008. www.theguardian.com.

Marr, Bernard. 2022. "How Shell Is Using Web3 and Blockchain for Sustainability and Energy Transition." *Forbes*, July 15, 2022. www.forbes.com.
Marx, Karl. 2004. *Capital: Volume I*. London: Penguin UK.
Marx, Karl, and Frederick Engels. 2010. *Marx & Engels Collected Works Vol 21: Marx and Engels; 1867–1870*. London: Lawrence & Wishart.
Marzocchi, Ottavio, and Martina Mazzini. 2022. *Pegasus and Surveillance Spyware*. Brussels: European Parliament. www.europarl.europa.eu.
Mascarenhas, Michael. 2012. *Where the Waters Divide: Neoliberalism, White Privilege, and Environmental Racism in Canada*. Lanham, MD: Lexington Books.
Matamoros-Fernández, Ariadna. 2017. "Platformed Racism: The Mediation and Circulation of an Australian Race-Based Controversy on Twitter, Facebook and Youtube." *Information, Communication & Society* 20 (6): 930–46.
Mateo, Juan José, Berta Ferrero, and Borja Andrino. 2022. "Madrid, el paraíso de la educación privada: En la capital son minoría los alumnos de la pública." *El País*, July 6, 2022.
Milei, Javier (@Jmilei). 2023. "We need to talk, Elon. . . ." *Twitter*, December 5, 2023, 11:43 p.m. https://twitter.com.
Millman, Oliver. 2024. "'Off the Charts': 2023 Was Hottest Year Ever Recorded Globally, US Scientists Confirm." *The Guardian*, January 12, 2024. www.theguardian.com.
Mawhinney, Sarah, Joey Reinhard, and Marni Lefebvre. 2023. *Tough Gig: Worker Perspectives on the Gig Economy*. Sydney, Australia: McKell Institute.
McGettigan, Andrew. 2013. *The Great University Gamble: Money, Markets and the Future of Higher Education*. London: Pluto Press.
McQuade, Brendan. 2019. *Pacifying the Homeland: Intelligence Fusion and Mass Supervision*. Berkeley: University of California Press.
McQuillan, Dan. 2022. *Resisting AI: An Anti-fascist Approach to Artificial Intelligence*. Bristol, UK: Policy Press.
Mejía Núñez, Gerardo. 2022. "La Blanquitud en México Según Cosas de Whitexicans." *Revista Mexicana de Sociología* 84 (3): 717–51.
Meta. 2022. "Meta Reports Fourth Quarter and Full Year 2021 Results." Press release, Meta Investor Realtions, February 2, 2022. https://investor.fb.com.
Metz, Laure, Jason E. Lewis, and Ludovic Slimak. 2023. "Bow-And-Arrow, Technology of the First Modern Humans in Europe 54,000 Years Ago at Mandrin, France." *Science Advances* 9 (8): eadd4675.
Mezzadra, Sandro, and Brett Neilson. 2013. *Border as Method, or, the Multiplication of Labor*. Durham, NC: Duke University Press.
Microsoft. 2019. "ExxonMobil to Increase Permian Profitability through Digital Partnership with Microsoft." *Microsoft News Center*, February 22, 2019. https://news.microsoft.com.
Mijente, Just Futures Law, and No Border Wall Coalition. 2021. *The Deadly Digital Border Wall*. N.p.: Mijente, Just Futures Law, and No Border Wall Coalition.
Milivojevic, Sanja. 2022. "Artificial Intelligence, Illegalised Mobility and Lucrative Alchemy of Border Utopia." *Criminology & Criminal Justice* 0 (0).

Miller, Carl. 2018. "Inside the British Army's Secret Information Warfare Machine." *Wired*, November 14, 2018.
Miller, Todd. 2019. *More Than a Wall: Corporate Profiteering and the Militarization of US Borders*. N.p.: Transnational Institute.
Milmo, Dan. 2021. "Rohingya Sue Facebook for £150bn over Myanmar Genocide." *The Guardian*, December 6, 2021.
Milmo, Dan, and Lisa O'Caroll. 2023. "Facebook Owner Meta Fined €1.2bn for Mishandling User Information." *The Guardian*, May 22, 2023. www.theguardian.com.
Mining Technology. 2024. "Chuquicamata Copper Mine." Accessed June 20, 2024. www.mining-technology.com.
Ministerio de Agricultura. 2023. "Fauna y flora nativa del Salar de Tara vuelve a tomar territorio." January 3, 2023. www.conaf.cl.
Molnar, Petra. 2020. *Technological Testing Grounds: Migration Management Experiments and Reflections from the Ground Up*. N.p.: Refugee Law Lab, EDRi.
Molnar, Petra. 2024. *The Walls Have Eyes: Surviving Migration in the Age of Artificial Intelligence*. New York: New Press.
Montoya, Ainhoa, Rachel Sieder, and Yacotzin Bravo-Espinosa. 2022. "Juridificación multiescalar frente a la industria minera: Experiencias de Centroamérica y México." *Íconos: Revista de Ciencias Sociales* 26 (72): 57–78.
Moore, Shannon Dawn Maree, Bruno De Oliveira Jayme, and Joanna Black. 2021. "Disaster Capitalism, Rampant Edtech Opportunism, and the Advancement of Online Learning in the Era of COVID19." *Critical Education* 12 (2): 1–21.
Morales, Héctor. 2013. "Construcción social de la etnicidad: Ego y alter en Atacama." *Estudios atacameños*, no. 46, 145–64.
Moreton-Robinson, Aileen. 2015. *The White Possessive: Property, Power, and Indigenous Sovereignty*. Minneapolis: University of Minnesota Press.
Moreton-Robinson, Aileen. 2019. "Terra Nullius and the Possessive Logic of Patriarchal Whiteness: Race and Law Matters." In *Changing Law: Rights, Regulation and Reconciliation*, edited by Marie Keyes, 123–36. London: Routledge.
Morozov, Evgeny. 2011. *The Net Delusion: How Not to Liberate the World*. London: Penguin.
Mould, Tom. 2020. *Overthrowing the Queen: Telling Stories of Welfare in America*. Bloomington: Indiana University Press.
Muench, Stefan, Eckhard Stoermer, Kathrine Jensen, Tommi Asikainen, Maurizio Salvi, and Fabiana Scapolo. 2022. *Towards a Green & Digital Future*. Luxembourg: Publications Office of the European Union.
Mulholland, Hélène. 2006. "Blair Welcomes Private Firms Into NHS." *The Guardian*, February 16, 2006. www.theguardian.com.
Mumford, Lewis. 2010. *Technics and Civilization*. Chicago: University of Chicago Press.
Muñiz, Ana. 2022. *Borderland Circuitry: Immigration Surveillance in the United States and Beyond*. Berkeley: University of California Press.
Munn, Nathan. 2018. "This Predictive Policing Company Compares Its Software to 'Broken Windows' Policing." *Vice Motherboard*, June 12, 2008. www.vice.com.

Muradov, Ibrahim. 2022. "The Russian Hybrid Warfare: The Cases of Ukraine and Georgia." *Defence Studies* 22 (2): 168–91.

Murray, Andrew. 2002. *Off the Rails: The Crisis on Britain's Railways*. London: Verso Books.

Muzzatti, Stephen L. 2022. "Strange Bedfellows: Austerity and Social Justice at the Neoliberal University." *Critical Criminology* 30 (3): 495–507.

Mytton, David. 2021. "Data Centre Water Consumption." *npj Clean Water* 4 (1): 11.

National Center for Education Statistics. n.d. "Number and Percentage of Undergraduate Students Enrolled in Distance Education or Online Classes and Degree Programs, by Selected Characteristics: Selected Years, 2003–04 through 2015–16." Accessed May 12, 2022. https://nces.ed.gov.

Ned Nemra, Casten. 2021. "Global Shipping Is a Big Emitter, the Industry Must Commit to Drastic Action Before it is Too Late." *The Guardian*, September 19, 2021. https://www.theguardian.com.

Nellemann, Christian, Rune Henriksen, Arnold Kreilhuber, Davyth Stewart, Maria Kotsovou, Patricia Raxter, Elizabeth Mrema, and Sam Barrat. 2016. *The Rise of Environmental Crime: A Growing Threat to Natural Resources, Peace, Development and Security*. Nairobi, Kenya: United Nations Environment Programme (UNEP).

Nemer, David. 2018. "The Three Types of Whatsapp Users Getting Brazil's Jair Bolsonaro Elected." *The Guardian*, October 25, 2018.

New Mexico Department of Justice. 2021. "Attorney General Hector Balderas Announces Landmark Settlements with Google Over Children's Online Privacy." Press release, December 13, 2021.

Neocleous, Mark. 2014. *War Power, Police Power*. Edinburgh, UK: Edinburgh University Press.

New Mexico *ex rel.* Balderas v. Google, LLC, 489 F. Supp. 3d 1254 (D.N.M. 2020) (No. 1:20-cv-00143-NF).

Ng, Abigail. 2019. "Smartphone Users are Waiting Longer before Upgrading—Here's Why." *CNBC*, May 17, 2019.

Nguyen, Aiha. 2021. *The Constant Boss: Work under Digital Surveillance*. N.p.: Data & Society. https://datasociety.net.

Noble, Safiya Umoja. 2018. *Algorithms of Oppression: How Search Engines Reinforce Racism*. New York: New York University Press.

Northcote Primary School. 2023. "Digital Technologies Policy (Internet, Social Media and Digital Devices)." Accessed June 20, 2024. www.northcoteps.vic.edu.au.

NYU. n.d. "Google Workspace Storage." Accessed Setptember 7, 2023. www.nyu.edu.

O'Neil, Cathy. 2016. *Weapons of Math Destruction: How Big Data Increases Inequality and Threatens Democracy*. New York: Crown.

Oberle, Bruno, Stefan Bringezu, Steve Hatfield-Dodds, Stefanie Hellweg, Heinz Schandl, and Jessica Clement. 2019. *Global Resources Outlook: 2019*. International Resource Panel, United Nations Envio. Nairobi, Kenya: United Nations Environment Programme.

Olivo, Antonio. 2023. "Northern Va. Is the Heart of the Internet. Not Everyone is Happy About That." *Washington Post*, February 10, 2023. www.washingtonpost.com.

Okpaleke, Francis N., Bernard Ugochukwu Nwosu, Chukwuma Rowland Okoli, and Ezenwa E. Olumba. 2023. "The Case for Drones in Counter-Insurgency Operations in West African Sahel." *African Security Review* 32 (4): 351–67.

Orr, Will, Kathryn Henne, Ashlin Lee, Jenna Imad Harb, and Franz Carneiro Alphonso. 2023. "Necrocapitalism in the Gig Economy: The Case of Platform Food Couriers in Australia." *Antipode* 55 (1): 200–221.

Ovetz, Robert 2022. "A Workers' Inquiry into Canvas and Zoom: Disrupting the Algorithmic University." In *Digital Platforms and Algorithmic Subjectivities*, edited by Emiliana Armano, Marco Briziarelli and Elisabetta Risi, 183–94. London: University of Westminster Press.

Ovetz, Robert. 2021. "The Algorithmic University: On-Line Education, Learning Management Systems, and The Struggle Over Academic Labor." *Critical Sociology* 47 (7–8): 1065–84.

Owen, John R., Deanna Kemp, Alex M. Lechner, Jill Harris, Ruilian Zhang, and Éléonore Lèbre. 2022. "Energy Transition Minerals and Their Intersection with Land-Connected Peoples." *Nature Sustainability* 6 (2): 203–11.

Pashukanis, Evgeny B. 1983. *Law and Marxism: A General Theory*. London: Pluto Press.

Pasquale, Frank. 2015. *The Black Box Society: The Secret Algorithms that Control Money and Information*. Cambridge, MA: Harvard University Press.

Pasquale, Frank. 2020. *New Laws of Robotics: Defending Human Expertise in the Age of AI*. Cambridge, MA: Harvard University Press.

Pasquinelli, Matteo. 2015. "Italian Operaismo and the Information Machine." *Theory, Culture & Society* 32 (3): 49–68.

Pasternak, Shiri, Deborah Cowen, Robert Clifford, Tiffany Joseph, Dayna Nadine Scott, Anne Spice, and Heidi Kiiwetinepinesiik Stark. 2023. "Infrastructure, Jurisdiction, Extractivism: Keywords for Decolonizing Geographies." *Political Geography* 101:102763.

Patel, Nilay. 2013. "Yes, Gmail Users Have an Expectation of Privacy." *The Verge*, August 15, 2013.

Pearce, Frank. 1976. *Crimes of the Powerful: Marxism, Crime and Deviance*. London: Pluto Press.

Pellegrini, Lorenzo, Murat Arsel, Martí Orta-Martínez, and Carlos F. Mena. 2020. "International Investment Agreements, Human Rights, and Environmental Justice: The Texaco/Chevron Case from the Ecuadorian Amazon." *Journal of International Economic Law* 23 (2): 455–68.

Pérez Gómez, Elena. 2021. "La Seca: Muerte silenciosa de la dehesa." *El Salto*, August 5, 2021. https://www.elsaltodiario.com.

Perrotta, Carlo, Kalervo N. Gulson, Ben Williamson, and Kevin Witzenberger. 2021. "Automation, Apis and the Distributed Labour of Platform Pedagogies in Google Classroom." *Critical Studies in Education* 62 (1): 97–113.

Pitarch, Sergi, and Lucas Marco. 2019. "La mayor sentencia colectiva contra Deliveroo considera a 97 'riders' empleados y no autónomos." *El Diario*, June 27, 2019. www.eldiario.es.

Phippen, William. 2021. "'A $10-Million Scarecrow': The Quest for the Perfect 'Smart Wall.'" *Politico*, December 10, 2021.

Phan, Thao, Jake Goldenfein, Monique Mann, and Declan Kuch. 2022. "Economies of Virtue: The Circulation of 'Ethics' in Big Tech." *Science as culture* 31 (1): 121–35.

Pintos Cubo, Julio César. 2022. "Proyecto de mina de Cañaveral: Una amenaza letal para acuíferos esenciales." *El Salto*, November 9, 2022. https://www.elsaltodiario.com.

Porter, Amanda. 2016. "Decolonizing Policing: Indigenous Patrols, Counter-Policing and Safety." *Theoretical Criminology* 20 (4): 548–65.

Porter, Amanda, and Chris Cunneen. 2020. "Policing Settler Colonial Societies." In *Australian Policing, Critical Issues in 21st Century Police Practice*, edited by Philip Birch, Michael Kennedy, and Erin Kruger, 97–411. London: Routledge.

Pozo Marín, Alejandro, and Ainhoa Ruiz Benedicto. 2022. *Combat Proven Business: Exporting The 'Israel Brand' to Maintain the Occupation and Normalise Injustice.* Barcelona: Centre d'Estudis per la Pau J. M. Delàs.

Prat, Felipe Díez, and Rubén Ranz Martín. 2020. "Mi experiencia como repartidor de Deliveroo y el intento por articular nuestra lucha desde la estructura sindical de UGT." *Teknokultura* 17 (2): 187–93.

Privacy International. 2013. *Eyes Wide Open: Special Report.* N.p.: Privacy International. https://privacyinternational.org.

Prokop, Andrew. 2023. "The Conservative Push for 'School Choice' Has Had its Most Successful Year Ever." *Vox*, September 11, 2023. https://www.vox.com.

Pueyo Busquets, Jordi. 2022. "Barcelona crea un 'software' para las escuelas alternativo a Google y a Microsoft." *El País*, February 9, 2022. https://elpais.com.

Quijano, Anibal. 2000. "Coloniality of Power and Eurocentrism in Latin America." *International Sociology* 15 (2): 215–32.

Rajagopal, Balakrishnan. 2003. *International Law from Below: Development, Social Movements and Third World Resistance.* Cambridge: Cambridge University Press.

Ramirez, Mary Kreiner. 2005. "The Science Fiction of Corporate Criminal Liability: Containing the Machine through the Corporate Death Penalty." *Arizona Law Review* 47 (4): 933–1002.

Redden, Joanna, Jessica Brand, and Vanesa Terzieva. 2020. "Data Harm Record (Updated)." Data Justice Lab. Last updated August 2020. https://datajusticelab.org.

Regulation (EU) 2022/2065 of the European Parliament and of the Council of 19 October 2022 on a Single Market for Digital Services and Amending Directive 2000/31/EC (Digital Services Act) (Text with EEA Relevance).

Reid, Alison, Elena Ronda-Perez, and Marc B. Schenker. 2021. "Migrant Workers, Essential Work, and COVID-19." *American Journal of Industrial Medicine* 64 (2): 73–77.

Restrepo, Eduardo. 2020. "Sujeto de la nación y otrerización." *Tabula Rasa*, no. 34, 270–88.

Richardson, Rashida. 2021. "Defining and Demystifying Automated Decision Systems." *Maryland Law Review* 81 (3): 785–840.
Richardson, Rashida, Jason M. Schultz, and Kate Crawford. 2019. "Dirty Data, Bad Predictions: How Civil Rights Violations Impact Police Data, Predictive Policing Systems, and Justice." *New York University Law Review Online* 94 (2019): 15–55.
Rigakos, George S. 2016. *Security/Capital: A General Theory of Pacification*. Edinburgh, UK: Edinburgh University Press.
Rio Tinto. 2021. "Rio Tinto Launches START: The First Sustainability Label for Aluminium." February 3, 2021. https://www.riotinto.com.
Rio Tinto. 2022. "Annual Report 2021." Accessed May 26, 2023. https://www.riotinto.com.
Rio Tinto. n.d. "Growth." Accessed May 26, 2023. https://www.riotinto.com.
Riofrancos, Thea. 2020. *Resource Radicals: From Petro-Nationalism to Post-Extractivism in Ecuador*. Durham, NC: Duke University Press.
Robinson, Bradley. 2021. "The ClassDojo App: Training in the Art of Dividuation." *International Journal of Qualitative Studies in Education* 34 (7): 598–612.
Robinson, Cedric J. 2020. *Black Marxism, Revised and Updated Third Edition: The Making of the Black Radical Tradition*. Chapell Hill: University of North Carolina Press Books.
Robinson, Mike, Kevin Jones, and Holger Janicke. 2015. "Cyber Warfare: Issues and Challenges." *Computers & Security* 49: 70–94.
Roborder. 2022. "Autonomous Swarm of Heterogeneous RObots for BORDER Surveillance." Accessed October 23, 2023. https://cordis.europa.eu.
Rodríguez, Dylan. 2020. *White Reconstruction: Domestic Warfare and the Logics of Genocide*. New York: Fordham University Press.
Rogoway, Mike. 2023. "Amazon Fuel Cells Would Use Natural Gas to Power Oregon Data Centers, Increasing Carbon Footprint." *The Oregonian*, February 6, 2023. https://www.oregonlive.com.
Romano, Valentina. 2023. "Parliament Adds Ecocide to EU's Draft List of Environmental Crimes." *Euractiv*, March 30, 2023. https://www.euractiv.com.
Romero, Amanda, José Aylwin, and Marcel Didier. 2021. *Globalización de las empresas de energía renovable: Extracción de litio y derechos de los pueblos Indígenas en Argentina, Bolivia y Chile*. Santigo, Chile: Centro de Información sobre Empresas y Derechos Humanos.
Rosenberg, Jacob. 2020. "'It's Feast or Famine': Gig Workers Bear the Risks of Coronavirus Because They Don't Have a Choice." *Mother Jones*, March 14, 2020.
Rosenthal, Caitlin. 2018. *Accounting for Slavery: Masters and Management*. Cambridge, MA: Harvard University Press.
Rossini, Patricia, Érica Anita Baptista, Vanessa Veiga de Oliveira, and Jennifer Stromer-Galley. 2021. "Digital Media Landscape in Brazil: Political (Mis) Information and Participation on Facebook and Whatsapp." *Journal of Quantitative Description: Digital Media* 1:1–27.
Rotaru, Victor, Yucheng Huang, Tianxiang Li, James Evans, and Ishanu Chattopadhyay. 2022. "Event-Level Prediction of Urban Crime Reveals a Signature of Enforcement Bias in US Cities." *Nature Human Behaviour* 6 (8): 1056–68.

Rushe, Dominic. 2013. "Google: Don't Expect Privacy When Sending to Gmail." *The Guardian*, August 15, 2013. www.theguardian.com.

Sa'di, Ahmad H. 2021. "Israel's Settler-Colonialism as a Global Security Paradigm." *Race & Class* 63 (2): 21–37.

Sadowski, Jathan, Salomé Viljoen, and Meredith Whittaker. 2021. "Everyone Should Decide How Their Digital Data Are Used—Not Just Tech Companies." *Nature* 595 (7866): 169–71.

Sainato, Michael. 2023. "'We Were Not Consulted': Native Americans Fight Lithium Mine on Site of 1865 Massacre." *The Guardian*, October 13, 2023. https://www.theguardian.com.

Saner, Emine. 2020. "Delivery Disaster: The Hidden Environmental Cost of Your Online Shopping." *The Guardian*, February 17, 2020. https://www.theguardian.com.

Sanger, David E. 2017. "Cyber, Drones and Secrecy." In *Understanding Cyber Conflict: 14 Analogies*, edited by George Perkovich and Ariel E Levite, 61–80. Washington, DC: Georgetown University Press.

Saunders, Daniel B. 2010. "Neoliberal Ideology and Public Higher Education in the United States." *Journal for Critical Education Policy Studies* 8 (1): 41–77.

Saunders, James, Priscillia Hunt, and John S. Hollywood. 2016. "Predictions Put into Practice: A Quasi-Experimental Evaluation of Chicago's Predictive Policing Pilot." *Journal of Experimental Criminology* 12 (3): 347–71.

Saura, Geo. 2016 "Neoliberalización filantrópica y nuevas formas de privatización educativa: La red global Teach for All en España." *Revista de Sociología de la Educación-RASE* 9 (2): 248–64.

Saura, Geo, Ekaitz Cancela, and Jordi Adell. 2022. "New Keynesianism or Smart Austerity? Digital Technologies and Educational Privatization Post COVID-19." *Education Policy Analysis Archives* 30:116.

Schaeffer, Felicity Amaya. 2022. *Unsettled Borders: The Militarized Science of Surveillance on Sacred Indigenous Land*. Durham, NC: Duke University Press.

Schavelzon, Salvador. 2012. *El Nacimiento del Estado Plurinacional de Bolivia, etnografía de una asamblea constituyente*. Santa Cruz de la Sierra, Bolivia: Clacso.

Schmelzer, Matthias, Andrea Vetter, and Aaron Vansintjan. 2022. *The Future Is Degrowth: A Guide to a World Beyond Capitalism*. London: Verso Books.

Schmitt, Michael N., ed. 2013. *Tallinn Manual on the International Law Applicable to Cyber Warfare*. Cambridge: Cambridge University Press, 2013.

Scholz, Jason, and Jai Galliott. 2018. "AI in Weapons: The Moral Imperative for Minimally-Just Autonomy." *US Air Force Journal of Indo-Pacific Affairs* 1 (2): 57–67.

Schrom, Ben. 2014. "Announcing Drive for Education: The 21st Century Backpack for Students." *Google Cloud Official Blog* (blog), September 30, 2014. https://cloud.googleblog.com.

Schwarz, Ori. 2019. "Facebook Rules: Structures of Governance in Digital Capitalism and the Control of Generalized Social Capital." *Theory, Culture & Society* 36 (4): 117–41.

Scott-Railton, John, Elly Campo, Bill Marczak, Bahr Abdul Razzak, Siena Anstis, Gizem Böcü, and Ronald Deibert. 2022. *CatalanGate: Extensive Mercenary Spyware Operation against Catalans Using Pegasus and Candiru*. Toronto: CitizenLab Report, University of Toronto.

Scott, Dayna Nadine. 2021. "Extractivism: Socio-Legal Approaches to Relations with Lands and Resources." In *The Routledge Handbook of Law and Society*, edited by Mariana Valverde, Kamari M. Clarke, Eve Darian Smith, and Prabha Kotiswaran, 124–27. London: Routledge.

Senor, Dan, and Saul Singer. 2011. *Start-Up Nation: The Story of Israel's Economic Miracle*. New York: McClelland & Stewart.

Shemitz, Leigh, and Paul Anastas. 2020. "Yale Experts Explain Microplastics." Yale Sustainability, December 1, 2020. https://sustainability.yale.edu.

Shilliam, Robbie. 2018. *Race and the Undeserving Poor: From Abolition to Brexit*. New Castle: Agenda Publishing.

Short, Damien, and Martin Crook, eds. 2022. *The Genocide-Ecocide Nexus*. London: Routledge.

Sibelco. 2023. "Sibelco Spruce Pine IOTA HPQ." YouTube video, 00:05:39, December 14, 2023. www.youtube.com/watch?v=MZf3ksYsHz8.

Siapera, Eugenia, and Paloma Viejo-Otero. 2021. "Governing Hate: Facebook and Digital Racism." *Television & New Media* 22 (2): 112–30.

Sierra, María Teresa. 2017. "Guerrero, Mexico: Community Police Confront Macro-Violences." *NACLA Report on the Americas* 49 (3): 366–69.

Simonite, Tom. 2021. "The Pentagon Scrubs a Cloud Deal and Looks to Add More AI." *Wired*, July 6, 2021. www.wired.com.

Simpson, Leanne Betasamosake. 2016. "Indigenous Resurgence and Co-Resistance." *Critical Ethnic Studies* 2 (2): 19–34.

Simpson, Leanne Betasamosake. 2017. *As We Have Always Done: Indigenous Freedom Through Radical Resistance*. Minneapolis: Univeristy of Minnesota Press.

Singer, Natasha. 2017. "How Google Took Over the Classroom." *New York Times*, May 13, 2017. https://www.nytimes.com.

Singh, Pawan. 2021. "Aadhaar and Data Privacy: Biometric Identification and Anxieties of Recognition in India." *Information, Communication & Society* 24 (7): 978–93.

Smalls, Christian, Angelika Maldonado, and Michelle Valentin Nieves. 2022. "The Workers Behind Amazon's Historic First Union Explain How They Did It." *Jacobin*, April 24, 2022. https://jacobin.com.

Soper, Spencer, Michael Tobin, and Michael Smith. 2021. "'Keep Driving': Amazon Dispatcher Texts Show Chaos amid Twisters." *Bloomberg Law*, December 17, 2021.

Sorochan, Cayley. 2012. "The Québec Student Strike—A Chronology." *Theory & Event* 15 (3).

Soto Aliaga, Nuria. 2023. *Riders on the Storm: Trabajadores de Plataformas de Delivery en Lucha*. Berlin: Rosa Luxemburg Stiftung. https://www.rosalux.eu.

Spice, Anne. 2018. "Fighting Invasive Infrastructures: Indigenous Relations against Pipelines." *Environment and Society* 9 (1): 40–56.

Srnicek, Nick. 2017. *Platform Capitalism*. Hoboken: John Wiley & Sons.
State of California Department of Justice. n.d. "CalGang® Frequently Asked Questions (FAQs)." Accessed June 29, 2022. https://oag.ca.gov.
Sterman, David. 2022. "Endless War Challenges Analysis of Drone Strike Effectiveness." *Journal of National Security Law and Policy* 13 (2): 305–18.
Stewart, Heather. 2022. "'I Just Want to Live': How UK Amazon Workers Came to Brink of Strike." *The Guardian*, October 12, 2022. www.theguardian.com.
Stewart, Heather. 2023. "Amazon Workers in Coventry to Request Union Recognition after Membership Doubles." *The Guardian*, April 26, 2023. www.theguardian.com.
Stony Brook. 2014. "Holy Cow! Unlimited Storage Coming to Google Drive." October 2, 2014. https://you.stonybrook.edu.
Stop Ecocide. 2021. "Legal Definition of Ecocide." Accesed June 2, 2024. www.stopecocide.earth.
Stop LAPD Spying Coalition. 2021. "Automating Banishment: The Surveillance and Policing of a Looted Land." November 9, 2021. https://automatingbanishment.org.
Stroud, Matt. 2021. "Heat Listed." *The Verge*, May 24, 2021.
Supran, Geoffrey, Stefan Rahmstorf, and Naomi Oreskes. 2023. "Assessing ExxonMobil's Global Warming Projections." *Science* 379 (6628): eabk0063.
Sutherland, Edwin H. 1945. "Is 'White Collar Crime' Crime?" *American Sociological Review* 1 (2): 132–39.
Svampa, Maristella. 2019. *Las fronteras del neoextractivismo en América Latina: Conflictos socioambientales, giro ecoterritorial y nuevas dependencias*. Bielefeld: Bielefeld University Press.
Tankosić, Ana, and Sender Dovchin. 2021. "(C)overt Linguistic Racism: Eastern-European Background Immigrant Women in the Australian Workplace." *Ethnicities* 23 (5): 726–57.
Tapia, Luis. 2007. "Una reflexión sobre la idea de estado plurinacional." *Osal* 22 (63): 47–64.
Taplin, Jonathan. 2017. *Move Fast and Break Things: How Facebook, Google, and Amazon Have Cornered Culture and What It Means for All of Us*. London: Pan Macmillan.
Tarnoff, Ben. 2022. *Internet for the People: The Fight for Our Digital Future*. London: Verso Books.
Tauri, Juan Marcellus, and Ngati Porou. 2014. "Criminal Justice as a Colonial Project in Settler-Colonialism." *African Journal of Criminology & Justice Studies* 8 (1): 20–37.
Tawil-Souri, Helga. 2012. "Digital Occupation: Gaza's High-Tech Enclosure." *Journal of Palestine Studies* 41 (2): 27–43.
Te Awekotuku, Ngahuia, and Linda Waimarie Nikora. 2003. *Nga taonga o te Urewera*. A Report prepared for the Waitangi Tribunal's Urewera District Inquiry (Wai 894, docB6).
Tiqqun. 2020. *The Cybernetic Hypothesis*. Cambridge, MA: MIT Press.
Tombs, Steve. 2016. "What to Do with the Harmful Corporation?" *Justice, Power and Resistance* 1:193–216.

Tombs, Steve, and David Whyte. 2015. *The Corporate Criminal: Why Corporations Must Be Abolished*. London: Routledge.
Tonnard, Yvon M. 2011. "Getting an Issue on the Table: A Pragma-Dialectical Study of Presentational Choices in Confrontational Strategic Maneuvering in Dutch Parliamentary Debate." PhD diss., University of Amsterdam.
Trade Union Congress. 2020. *Challenging Amazon: What Can We Do about Amazon's Treatment of Its Workers?* London: TUC.
Transport Workers' Union. 2020. "Food Delivery Rider/Driver Survey." January 24, 2020. https://www.twu.com.au.
Trautman, Lawrence J., and Peter C. Ormerod. 2017. "Industrial Cyber Vulnerabilities: Lessons from Stuxnet and the Internet of Things." *University of Miami Law Review* 72 (3): 761–826.
Tronti, Mario. 2019. *Workers and Capital*. London: Verso Books.
Troupe, Mauvaise. 2018. *The Zad and NoTAV: Territorial Struggles and the Making of a New Political Intelligence*. London. Verso Books.
Tzouvala, Ntina. 2020. *Capitalism as Civilisation: A History of International Law*. Cambridge: Cambridge University Press.
Uber Eats. 2023. "Get Almost Almost Anything." YouTube Video, 00:01:00, September 15, 2023. www.youtube.com/watch?v=n5r4c6ru_jo.
Uber. 2018. "Doors Are Always Opening." YouTube Video, 00:00:36, September 18, 2018. www.youtube.com/watch?v=kauVGsP5xMY.
Uber. 2020. "Driving a Grean Recovery." September 8, 2020. www.uber.com.
Uber. 2022. "Do Less." YouTube Video, 00:01:00, October 10, 2018. www.youtube.com/watch?v=Dqv8ppQIoNw.
UN Environment Programme. 2018. "Gaps in International Environmental Law and Environment-Related Instruments: Towards a Global Pact for the Environment—Report of the Secretary-General." Report for the United Nations General Assembley, 73rd session, November 30, 2018. https://wedocs.unep.org.
UN Human Rights Council. 2019. *Annex to the Report of the Special Rapporteur on Extrajudicial, Summary or Arbitrary Executions: Investigation into the Unlawful Death of Mr. Jamal Khashoggi*. Geneva: United Nations.
UNECE. 2022. "Ensuring a Sustainable Supply of Critical Raw Materials Is Essential for Achieving Low-Carbon Energy Transition." Press release, November 9, 2022. https://unece.org.
United Nations. 2007. *United Nations Declaration on the Rights of Indigenous Peoples*. N.p.: United Nations. www.ohchr.org/.
United Nations. 2019. *The Sustainable Development Goals Report 2019*. New York: United Nations. https://digitallibrary.un.org.
University of Queensland. 2024. "Calculate Your H-Index. Learn about the H-Index and How It Is Used to Measure Research Output." May 31, 2024. https://guides.library.uq.edu.au.
Urbán, Miguel. 2023. *Trumpismos: Neoliberales y autoritarios—Radiografía de la derecha radical*. Barcelona: Verso Libros.

US Department of State. 2023. "Political Declaration on Responsible Military Use of Artificial Intelligence and Autonomy." November 9, 2023. www.state.gov.

US Census. 2021. "Selected Economic Characteristics." Accessed January 10, 2024. https://www.census.gov.

US Geological Survey. 2023. *Mineral Commodity Summaries 2023*. N.p.: US Geological Survey. https://pubs.usgs.gov.

Valdivia, Ana, and Martina Tazzioli. 2023. "Datafication Genealogies beyond Algorithmic Fairness: Making Up Racialised Subjects." In *FAccT '23: Proceedings of the 2023 ACM Conference on Fairness, Accountability, and Transparency*, 840–50. New York: Association for Computing Machinery.

Valtysson, Bjarki. 2022. "The Platformisation of Culture: Challenges to Cultural Policy." *International Journal of Cultural Policy* 28 (7): 786–98.

Van Bekkum, Marvin, and Frederik Zuiderveen Borgesius. 2021. "Digital Welfare Fraud Detection and the Dutch SyRI Judgment." *European Journal of Social Security* 23 (4): 323–40.

Van den Berg, Stephanie. 2021. "Dutch Government Quits over 'Colossal Stain' of Tax Subsidy Scandal." *Reuters*, January 15, 2021. www.reuters.com.

Van Dijck, José. 2014. "Datafication, Dataism and Dataveillance: Big Data between Scientific Paradigm and Ideology." *Surveillance & Society* 12 (2): 197–208.

van Veen, Christiaan. 2020. "Landmark Judgment from the Netherlands on Digital Welfare States and Human Rights." *Open Global Rights*, March 19, 2020. www.openglobalrights.org.

Venier, Silvia. 2019. "The Role of Facebook in the Persecution of the Rohingya Minority in Myanmar: Issues of Accountability under International Law." *Italian Yearbook of International Law Online* 28 (1): 231–48.

Vervloesem, Koen. 2020. "How Dutch Activists Got an Invasive Fraud Detection Algorithm Banned." Algorithm Watch. Accesed April 2, 2024. https://automatingsociety.algorithmwatch.org.

Vidalou, Jean-Baptiste. 2017. *Être forêts: Habiter des territoires en lutte*. Paris: Zones.

Vilca, Tomás, Lidia Reyes Berna, Esmeralda Ramos, Minette Zuleta Mondaca, Alejandra Varas Mora, Margarita Chocobar Cruz, Gina Cruz. 2022. *Camino ancestral del arte rupestre Lickanantay*. Santiago, Chile: Ministerio de las Culturas las Artes y el Patrimonio.

Virno, Paolo. 2003. *A Grammar of the Multitude*. London: Semiotext(e).

Vitale, Alex S. 2021. *The End of Policing*. London: Verso Books.

Von der Leyen, Ursula. 2021a. "Speech by President Von der Leyen at the Global Leaders Summit Hosted by U.S. President Joe Biden on the Occasion of Earth Day." European Commission. April 22, 2021. https://ec.europa.eu.

Von der Leyen, Ursula. 2021b. "State of the Union Address by President Von Der Leyen." European Commission. September 15, 2021. https://ec.europa.eu.

Von der Leyen, Ursula. 2022. "State of the Union Address by President Von Der Leyen." European Commission. September 14, 2022. https://ec.europa.eu.

Wacquant, Loïc. 2009. *Punishing the Poor: The Neoliberal Government of Social Insecurity*. Durham, NC: Duke University Press.

Waldman, Ari Ezra. 2019. "Power, Process, and Automated Decision-Making." *Fordham Law Review* 88 (2): 613–32.
Walton, Allan, Paul Anderson, Gavin Harper, Vicky Mann, John Beddington, Andy Abbott, Andrew Bloodworth et al. 2021. *Securing Technology-Critical Metals for Britain*. Birmingham: University of Birmingham. 2021.
Wang, Jackie. 2018. *Carceral Capitalism*. Cambridge, MA: MIT Press.
Wang, Nicole, Alvaro M. Bedoya, Daniel Bateyko, and Elizabeth Tucker. 2022. "American Dragnet: Data-Driven Deportation in the 21st Century." N.p.: Center on Privacy and Technology at Georgetown Law.
Wang, Xuefeng, and John Tomaney. 2019. "Zhengzhou—Political Economy of an Emerging Chinese Megacity." *Cities* 84:104–11.
Ward, Annie. 2022. "Ask a Techspert: What's a Subsea Cable?" *Google Key Word*, January 28, 2022. https://blog.google.
Warofka, Alex. 2018. "An Independent Assessment of the Human Rights Impact of Facebook in Myanmar." *Facebook Newsroom*, November 5, 2018.
Warwick Valley Central School District. 2021. *Chromebook Policy Handbook 2021–22*. N.p.: Warwick Valley Central School District. https://www.warwickvalleyschools.com.
Watego, Chelsea. 2021. "'Who Are the Real Criminals?': Making the Case for Abolishing Criminology." 43rd John Barry Memorial Lecture, University of Melbourne, November 29, 2021.
Watson, Andrew J., Ute Schuster, Jamie D. Shutler, Thomas Holding, Ian GC Ashton, Peter Landschützer, David K. Woolf, and Lonneke Goddijn-Murphy. 2020. "Revised Estimates of Ocean-Atmosphere CO_2 Flux are Consistent with Ocean Carbon Inventory." *Nature Communications* 11 (1): 4422.
Watson, Irene. 2018. "Aboriginal Relationships to the Natural World: Colonial 'Protection'of Human Rights and the Environment." *Journal of Human Rights and the Environment* 9 (2): 119–40.
Weil, Patrick. 2009. "Why the French Laïcité Is Liberal." *Cardozo Law Review* 30 (6): 2699–714.
Weis, Valeria Vegh. 2019. "Towards a Critical Green Southern Criminology: An Analysis of Criminal Selectivity, Indigenous Peoples and Green Harms in Argentina." *International Journal for Crime, Justice and Social Democracy* 8 (3): 38–55.
White House. 2023. "Fact Sheet: President Biden's Budget Strengthens Border Security, Enhances Legal Pathways, and And Provides Resources to Enforce our Immigration Laws." March 9, 2023. https://www.whitehouse.gov.
Whiteford, Peter. 2021. "Debt by Design: The Anatomy of a Social Policy Fiasco—Or Was It Something Worse?" *Australian Journal of Public Administration* 80 (2): 340–60.
Whyte, David. 2014. "Regimes of Permission and State-Corporate Crime." *State Crime Journal* 3 (2): 237–46.
Whyte, David. 2019. "Death to the Corporation: A Modest Proposal." In *A World Turned Upside Down*, edited by Leo Panitch and Greg Albo, 289–304. New York: Monthly Review Press.

Whyte, David. 2020. *Ecocide: Kill the Corporation Before It Kills Us*. Manchester, UK: Manchester University Press.

Williamson, Ben. 2017. "Decoding Classdojo: Psycho-Policy, Social-Emotional Learning and Persuasive Educational Technologies." *Learning, Media and Technology* 42 (4): 440–53.

Williamson, Ben, and Anna Hogan. 2020. *Commercialisation and Privatisation in/of Education in the Context of Covid-19*. Brussels: Education International.

Wilmshurst, Janet. 2007. "Story: Human Effects on the Environment." Te Ara—the Encyclopedia of New Zealand. Accessed April 5, 2023. http://www.TeAra.govt.nz.

Wolfe, Patrick. 2006. "Settler Colonialism and the Elimination of the Native." *Journal of Genocide Research* 8 (4): 387–409.

Wolkmer, Antonio Carlos. 2019. *Teoría crítica del derecho desde América Latina*. Madrid: Ediciones Akal.

World Bank Data. n.d. "Individuals Using the Internet (% of Population)—2022." Accessed December 12, 2023. https://data.worldbank.org.

World Integrated Trade Solution. 2021. "Carbonates: Lithium Carbonate Exports by Country in 2021." Accessed May 12, 2023. https://wits.worldbank.org.

Xnet. 2022. "Introducing DD: A Tool for the Democratic Digitalisation of Education." *Xnet*, February 9, 2022. https://xnet-x.net/en.

Yeung, Karen. 2018. "Algorithmic Regulation: A Critical Interrogation." *Regulation & Governance* 12 (4): 505–23.

Yu, Jun, and Nick Couldry. 2022. "Education as a Domain of Natural Data Extraction: Analysing Corporate Discourse about Educational Tracking." *Information, Communication & Society* 25 (1): 127–44.

Yue, Neriah. 2020. "The 'Weaponization' Of Facebook in Myanmar: A Case for Corporate Criminal Liability." *Hastings Law Journal* 71 (3): 813–44.

Zhou, Naaman. 2020. "Australia's Delivery Deaths: The Riders Who Never Made it and the Families Left Behind." *The Guardian*, November 21, 2020. https://www.theguardian.com.

Zierler, David. 2011. *The Invention of Ecocide: Agent Orange, Vietnam, and the Scientists Who Changed the Way We Think About the Environment*. Athens: University of Georgia Press.

Zuberi, Tukufu. 2001. *Thicker Than Blood: How Racial Statistics Lie*. Minneapolis: University of Minnesota Press.

Zuboff, Shoshana. 2019. *The Age of Surveillance Capitalism: The Fight for a Human Future at the New Frontier of Power*. London: Profile.

Zuckerberg, Mark. 2021. "I Wanted to Share a Note I Wrote to Everyone at Our Company." October 6, 2021. https://www.facebook.com.

Zureik, Elia. 2001. "Constructing Palestine through Surveillance Practices." *British Journal of Middle Eastern Studies* 28 (2): 205–27.

Zureik, Elia. 2020. "Settler Colonialism, Neoliberalism and Cyber Surveillance: The Case of Israel." *Middle East Critique* 29 (2): 219–35.

INDEX

Page numbers in italics indicate Figures

Aadhaar system, 34
abolitionism, 29, 42, 165, 198–99, 204
academia, 15, 18, 28, 51, 108–12, 118, 172; digitization of, 57–62; *h*-index and, 60–61
acceleration, 5–6, 10, 12, 162, 177, 183; of privatization, 28, 76, 83
access, accessibility and, 7, 12–13; education, 68, 72–74, 81–82; internet, 9, 56, 143, 145–46, 185
accountability, 13, 50, 107–8, 188, 191, 196, 204; education and, 67, 69–70
accumulation, capitalist, 41, 43, 119–20, 124–25, 138, 166, 201
adjunct faculty, 58–59
Advanced Recognition Technology System (HART system), DHS, 103
advertising, 8, 125–26, 133, 156
AeroVironment, 103
Afghanistan, 106
Africa, 100, 195–96
Agency for Criminal Justice, EU, 174
Agozino, Biko, 165
agriculture, 12, 179, 186–87
Agusdinata, Datu B., 180
Ahmed, Akbar, 108
Albanese, Anthony, 37
Albemarle, 178–80, 182–83
Alexandria-Portland Titan, 155
Algayerova, Olga, 158
Algeria, 120
"algorithmic fairness," 51–52

algorithmic justice, 13–14, 52
algorithms, 5–9, 21–22, 123, 190, 195–96, 198; bias, 31–32, 51; Chicago police department use of, 109–10; crime data, 110–11; decision-making and, 13; denial of services via, 50; domination by, 52; education, 78; errors and, 31–32, 37; exploitation and, 28, 115–17, 125–29; oppression and, 35–37, 45; as proprietary, 127, 129; racism and, 26–27, 43; surveillance, 86, 128
Alimonda, Hector, 177
Alphabet, 7, 12–13, 107, 188, 192, 197. *See also* Google
Alston, Philip, 50
Amazon, 7, 23, 113, 115, 117, 145–47, 150, 192; borders and, 88; carbon emissions, 8; deaths related to, 1–2; UK, 130–32; unionization efforts and, 130–31; warehouses, 1–2, 8, 55, 131–32; worker demographics, 125
American Dragnet Data-Driven Deportation in the 21st Century (Georgetown Center for Privacy and Technology), 103–4
Andreessen, Marc, 14
Andrejevic, Mark, 15, 26
Androids, Android operating system and, 7, 147, 185, 188, 192
Anduril Industries, 88, 103, 104
anticompetitive behavior, 6–7, 192
antiunion tactics, 117, 130–32

261

Aotearoa-New Zealand, 167–70, 178, 187
Apple, 144–47
aquifers, 134, 136–37, 180
ArcelorMittal, 155
Argentina, 138, 177–78, 182, 185
Arizona, 145–46
Artificial Intelligence (AI), 14, 51, 85–95, 112, 117, 194
Asia, 125, 185
Asociación de Medios de Información de España, 15
assassinations, 9, 105–8, 141
asymmetrical relationships, 8, 20, 117, 122–23, 128, 182
Atacameño People's Council (CPA), 179
Atiles, José, 165–66
Atlantic slave trade, 119–20. *See also* slavery
austerity, 45–47, 153
Australia, 15, 91, 106, 120–21, 135–37, 178, 192; Centrelink, 37; delivery workers in, 115–16, 123; education in, 73–75; fraud detection in, 35; Human Rights Commission, 51; patriarchy in, 42; welfare in, 68
automated decision-making system, 13, 27, 32–36, 50–52, 94
"Automating Inequality" (Eubanks), 47
automation, 12–13, 25, 32, 50, 52, 104, 154; border control, 102–3; debt recovery programs, 37; police and, 97–98; risk assessment and, 21, 101, 109–10

Baars, Grietje, 24, 199
Badische Anilin-und Sodafabrik (BASF SE), 155
Balderas, Hector, 81
Barcelona, Spain, 3, 11, 34–35, 54–55
Barzani, Jasmine, 165
BASF SE. *See* Badische Anilin-und Sodafabrik
batteries, 133, 159, 176–77
Benjamin, Ruha, 21, 43

Benjamin, Walter, 148
Benvegnù, Carlotta, 13
Bescherming Burgerrechten, 33
Betasamosake Simpson, Leanne, 166–67, 171
bias, 31–32, 51, 109–10, 185
Biden, Joe, 91–92
Big Tech, 6, 8, 15, 18, 21–22, 63; greenwashing by, 146–47; regulations, 193–98. *See also specific companies*
billionaires, 5, 9, 14, 54
biometric data, 34, 86, 102–3
Bittle, Steven, 199
Black Friday, 8
Black people, 38, 125
Blair, Tony, 69
blockchain, 153
Blumenthal, Richard, 191
Bolivia, 120, 138, 170, 177–78, 185, 201
Borderland Circuitry (Muñiz), 99
borders, 39, 88, 92, 100–104
Bourdieu, Pierre, 40
bourgeoisie, 24, 135, 206
Brantingham, Jeff, 110–11
Braverman, Harry, 61
Brayne, Sarah, 26
Brazil, 47, 149, 171, 185, 187–88
British Petroleum, 155
Broussard, Meredith, 21
"Brussels effect," 194
Bush, George W., 70

Cáceres, Spain, 135–37, 140, 147–48, 203–4
Caffentzis, George, 118
CalGang database, 104
"Caliban and the Witch" (Federici), 119
Cambridge Analytica scandal, 54, 97, 192
Campaña del Desierto (1878–90), Argentina, 178
Campbell, Kathryn, 37
Canada, 15, 101, 106, 120, 141, 145, 155, 178
Cancela, Ekaitz, 156
Capital (Marx), 122–23

capitalism, 6, 12, 14, 19, 29–30, 94, 201–2, 205; accumulation and, 41, 43, 119–20, 124–25, 138, 166, 201; fossil, 10–11; global, 4, 42, 88, 152–53; punitive power and, 198, 204; racial, 4, 39–42, 70, 97–98, 125–30; surveillance, 56, 59
Caplan, Jeremy, 62–63
carbon emissions, 8, 10, 147, 157, 172–73
cargo ships, 144
Caribbean, 125
cars, 12, 133, 147, 176–77
Castro-Gómez, Santiago, 38
Catalonia, Spain, 80–81
CBP. *See* Customs and Border Protection
cell phones. *See* mobile phones
censorship, 8–9, 90
Chamayou, Gregoire, 67
changes, technological, 28, 157, 176–77
charter schools, 71
Chicago, Illinois, 108–10
children, 86, 191; data of, 77–78, 188
Chile, 29, 146, 176–83, 185
China, 144, 149, 158, 178, 185, 194
Chromebooks, Google, 73–74, 77–78
Chun, Wendy, 38–39
Chuquicamata copper mine, 177
Citizen Lab, University of Toronto, 105
citizenship, 31–32, 44–45, 73, 154–55
civilians, 85–86, 90–92, 95, 97, 107, 112
civil wars, 20, 96, 155
classes, social, 20–22, 28, 96, 99, 108–11
Client Registry Information System-Enhanced (CRIS-E system), 46
Clifford, Robert, 203
Climate Accountability Institute, 10
climate change, 10–11, 152, 154, 158–59, 164
CO2 emissions, 8, 10, 147, 157, 172–73
cobalt, 158
code, computer, 123–24. *See also* algorithms
Codex Theodosianus, 123
collective action, 113–14, 203–6
collective rights, 14, 17, 171, 197

colonialism, colonization and, 23–24, 27–28, 38, 140–41, 158; Aotearoa-New Zealand, 167–70; capitalism and, 165–66, 202; data, 15, 25–26, 188; imperialism and, 93; Israel and, 87–88; racialization and, 44, 112–13; settler, 88, 99, 141; South American, 120, 170, 177–79; surveillance and, 88, 112–13
"color-blindness," 44, 109–10
Commission Nationale de l'Informatique et des Libertés, France, 192
commodification, 66–68, 149
computational power, 6, 15
computer language, 123–24
Congo, 158
Congressional Research Service, US, 6
consent, 15, 80, 104, 192
consumption, 10, 12, 23, 147–51, 162, 183
CoopCycle, 132
Cooperative Cyber Defense Center of Excellence, NATO, 92
copper mining, 177
CORFO. *See* Production Development Corporation
corporations, 6–7, 35, 95, 112, 125–28; criminality, 4–5, 8–10, 13, 16–26, 52–53, 100, 188–206; disinformation spread by, 96–97; impunity of, 20–21, 24; management, 67; power of, 4, 16, 18, 55, 77, 104, 117, 200; responsibilities of, 40, 51, 83, 188–203. *See also specific corporations*
Couldry, Nick, 15, 25–26
COVID-19 pandemic, 2–3, 8, 14, 55, 116
Cowen, Deborah, 141
CPA. *See* Atacameño People's Council
Crawford, Kate, 51
crime, 100, 108–11, 174–83
crimes, state and corporate, 4–5, 8–10, 13, 16–26, 52–53, 100
criminalization, 41, 99, 104
criminal justice, 36, 42, 43
criminal law, 20, 173, 199

criminology, 110–11, 165–67; critical, 4–5, 21–22, 189, 198, 204
CRIS-E system. *See* Client Registry Information System-Enhanced
critical criminology, 4–5, 21–22, 189, 198, 204
critical race studies, 118–20
Critical Raw Materials Act, EU, 133, 139, 160, 183
Cusicanqui, Silvia Rivera, 42
Customs and Border Protection (CBP), US, 100–103
cyber defenses, 15, 92, 111
cybernetics, 14, 102, 124
cyberspace, 91–92, 124
Cyberwar and Revolution (Matviyenko, Dyer-Witheford), 96
cyberwarfare, 19, 85–88, 101–4, 112–13; academia and, 108–11; dispossession and, 28, 96–100; military AI and, 89–95; surveillance and, 105–8

DARPA. *See* Defense Advanced Research Projects Agency
data, 12, 17, 86, 97, 108–11, 129, 192; biometric, 34, 86, 102–3; of children, 15, 77–81, 188; colonialism, 15, 25–26, 188; common ownership of, 205; mismanagement, 31–34; racism and, 22
databases, 31–32, 104
data centers, 92, 145–48
Data Protection Commission, Ireland, 192
Datatilsynet, 79
Davis, Angela, 40–41
Davis, Antigone, 191
deaths, 1–3, 85–88, 100, 107–8, 178; assassinations as, 9, 105–8, 141; factory, 155; premature, 41, 44. *See also* ecocide
decarbonization, 8, 12, 162, 176–77
decentralization, 107
decision-making, 13, 27, 32–36, 50–52, 94, 205
decolonization, decolonialism and, 27, 29, 120, 202

Deepwater Horizon oil spill, 155
Defense Advanced Research Projects Agency (DARPA), 97, 108, 110
defense contractors, 101, 103–4
deforestation, 151–52, 172–73 , 180
dehumanization, 95, 107, 130
Delfanti, Alessandro, 121
Deliveroo, 117, 124, 132
delivery workers, 1–3, 55, 115–16, 124, 127–30
Deloitte, 156–57
democracies, democratic values and, 14, 194, 197, 201, 205
democratization, 105, 205
Denmark, 79–80, 145
Department of Homeland Security (DHS), US, 101, 103
dependency, 71; fossil-fuel, 12, 23; technological, 4, 158, 187–88
depopulation, 133, 137
deportation, 22
depression, 18, 190
deprofessionalization, 32–33
DHS. *See* Department of Homeland Security
digital age, 4–5, 100, 104, 153
"digital border wall," 100–104
digital capitalism. *See specific topics*
digital citizens, 73
digital divide, 185
digital enclosures, 7, 15, 77
digital infrastructure, 9, 63, 65, 73, 75
digitalization, 5, 27, 75–76, 83, 101, 138, 154, 186, 206; of imperialism, 107–8
Digital Markets Act, EU, 194
Digital Services Act, EU, 194–97
digitized racial neoliberalism, 27, 35
Dijck, Jose Van, 13
Directive 2008/99/EC, EU, 174
discrimination, 13, 22, 31–32, 43–44, 101, 108–11
disinformation, 16–17, 96–97
dispossession, 25, 28, 95, 177–78, 182; cyberwarfare and, 96–100; genocide

and, 85–87; Indigenous, 42, 120, 138; neoliberal, 47; racialized, 41; working class, 62
drones, 88–92, 101, 107–8
droughts, 10–12, 134
drug trafficking, 21–22, 170–71
dual citizenship, 32
Dubal, Veena, 127–28
Du Bois, W. E. B., 42
due process, 80–81, 106–7
Durruti, Buenaventura, 206
Dutch Tax Agency, 31–34, 52
Dyer-Whiteford, Nick, 14, 96
Dynamic Models of Insurgent Activity project, 110–11

East India Company, 24–25
ecocide, 29, 147, 163, 166, 198–200, 202, 204; defining, 164, 172–76; lithium, 176–82
economic development, 11, 176, 180
Ecuador, 170, 173
EDRi. *See* European Digital Rights network
EdTech. *See* educational technology
Eduardo Bonilla-Silva, 44
education, educational institutions and, 27, 115–16, 202–3; accessibility and, 68, 72–74, 81–82; COVID-19 impacting, 55, 59, 72–73; educational infrastructures and, 13, 28, 56, 64, 66, 72, 83; infrastructure, 13, 28, 56, 64–66, 65, 72, 83; LMS technologies and, 62; online, 58, 61–62, 72–73; outsourcing in, 62–63, 71–72
educational technology (EdTech), 27–28, 65–66, 71–76
Edwardsville, Illinois, 1
Ekaitz Cancela, 26
Elbit Systems, Israeli, 19, 88, 103–4
electric cars, 12, 133, 147, 176–77
Electronic Frontier Foundation, 64
emails, college, 62–66, 65

enclosures, digital, 7, 15, 77
energy transition minerals, 23
Environmental Courts, Chile, 182
environmental destruction, 4, 6–7, 20, 23–24, 29, 153, 155, 174–82; CO_2 emissions and, 8, 10, 147, 157, 172–73; deforestation as, 151–52, 172–73 , 180. *See also* ecocide
E.ON, 155
"e-prison," 98
Ernst & Young, 156–57
ERT. *See* European Round Table of Industrialists
Escobar, Arturo, 177
essential workers, 3, 116
ethics, 6, 88, 188–89; AI weaponry and, 86–87, 93–94, 112
EU. *See* European Union
Eubanks, Virginia, 32, 47
eugenics, 38–39, 43
Eurocentrism, 41, 93–94, 112–13, 202
Europe, European and, 11–12, 22, 24–25, 42, 49, 92, 151, 157–62; on AI moderation, 51; Amazon centers, 8; courts, 48, 192, 194, 196; ROBORDER initiative, 102
European Commission, 154, 156–62, 175–76
European Digital Rights network (EDRi), 92
European Digital Services Act, 189
European Round Table of Industrialists (ERT), 155–56
European Union (EU), 10–11, 44–45, 142, 174–76, 185–86, 192, 194; Agency for Criminal Justice, 174; on automation, 32; borders and, 100–102; Digital Services Act, EU, 194–97; General Data Protection Regulation, 79; green and digital transition promoted by, 133–36, 153–61; on human rights, 160–61; racialization and, 120
Ever Alot (cargo ship), 144

e-waste, 23–24
exploitation, 5–7, 14, 22, 38, 42–45, 57–58, 121–22; algorithms and, 28, 115–17, 125–29; extractivism and, 10, 133–34, 162, 177–82
export models, 12, 19
expropriation, 122–23
extinction, 152, 172
extractivism, 11–12, 23, 25, 29, 133–40, 147, 170, 181–82; European Commission and, 158–62, *159*; in the Salar de Atacama, 176–80
Extremadura New Energies, 135–36, 138–39
ExxonMobil, 51

Facebook, 6–9, 15, 18, 29–30, 186–88, 197; Cambridge Analytica scandal, 54, 97, 192; Facebook Files, 189–93
facial recognition, 21–22, 86, 101–3
factories, 121–25, 133, 144, 155
fake news, 16, 17
Family Educational Rights and Privacy Act, US, 64
far-right movements and politics, 27, 47–48, *49*, 54
fascism, 6, 22, 155
Federal Trade Commission, US, 192–94
Federici, Silvia, 119
feminism, 118–19, 206
financial crisis (2008), 14, 17
financialization, 72, 82–83
fines, 19, 52, 192, 197
First Nations people. *See* Indigenous and First Nations people
"Five Eyes Alliance" (Canada, United Kingdom, Australia, and New Zealand), 106
Food and Agriculture Organization, 172
food delivery, 55, 115–16, 124
forests, 151–52, 167–70
fossil fuels, 8, 10–11, 23, 147, 154
Foucault, Michel, 38, 67

Foxconn plant, 144
France, 48, *49*, 120, 126
fraud, fraud detection and, 34–35, 45, 50
freelancers, 8, 57
free market, 7
Friedman, Milton, 68, 197
Friends of the Earth Europe, 160
Frontex (European Border and Coast Guard Agency), 92, 100
Fuchs, Christian, 26

Gago, Veronica, 67
Gajardo, Gonzalo, 180–81
Galston, Arthur W., 163
García Linera, Álvaro, 201, 203
Gaza, occupied, 8–9, 19, 86, 89–90, 93
GDP. *See* gross domestic product
Gebrialhas, Dalia, 125
Gebru, Timnit, 26
gender, 18, 21, 25, 109–10, 118–19, 130, 190
General Data Protection Regulation, 79, 194–95, 203
Generation Z., 118
genocide, 8–9, 85–86, 93, 163, 166, 169, 188
gentrification, 12, 54–55
George W., 70
Germany, 22, 39, *49*, 87, 155
Giannocopoulos, Maria, 165
gig workers, 3–4, 57, 115–19, 126
Gil, Juan, 137
Gilmore, Ruth Wilson, 40–41, 119–20
global capitalism, 4, 42, 88, 152–53
Global Footprint Network, 150–51
Global North, 7, 11–12, 14, 25, 162, 187, 189, 202; deforestation, 151; ecological movements in, 164–65; hierarchization by, 42; neoliberalism and, 45–46; privilege and, 120
Global South, 24–25, 42, 151–52, 161–62, 187
Global System for Mobile Communications (GSM) metadata, 106
global warming, 10–11, 134, 152
Glovo, 132

Goikoetxea, Jule, 119
Goldman Sachs, 176
González, Roberto J., 96–97
Google, 5, 64, 113, 145–46, 188, 192, 197; Caplan on, 62–63; impact on the field of education, 73–82; Stony Brook University and, 63–65, 65
"Google and Microsoft" (Caplan), 62–63
Google Scholar, 60
Gorz, André, 148
Gospel (AI tool), 85–86
governance, 14, 189, 194–98
Goya, Francisco de, 152
GPS tracking, 98
Gramsci, Antonio, 199
Great Depression, 169
Great Pacific Garbage Patch, 172
green and digital transition, 12, 16, 23, 29, 133–39, 147–48, 153–62
green criminology, 165–67
greenhouse gases, 10, 157–58
green imperialism, 157–62
Greenwald, Glen, 105
greenwashing, 14, 135–36, 145–48
Grosfoguel, Ramón, 42
gross domestic product (GDP), 149
growth, 140, 148–53, 154, 156–58
GSM. *See* Global System for Mobile Communications
Guardian (newspaper), 102, 105, 107, 131–32
Gulf of Mexico, 172–73
Guterres, António, 10

Haidt, Jonathan, 190
Haiti, 171
Hamas, 85
Harney, Stefano, 141
HART system. *See* Homeland Advanced Recognition Technology System
hate speech, 188, 196
Haugen, Frances, 18, 198–91
health care, 3, 69, 121, 126, 202–3
Helsingør, Denmark, 79–80

Hermes UAVs, 19
Higgins, Polly, 164, 175
Hildebrandt, Mireille, 36
Hillyard, Paddy, 199
h-index, 60–61
Hitler, Adolf, 22
Holland, 34
Homeland Advanced Recognition Technology System (HART system), DHS, 103
Hong Kong, 105
Horne, Gerald, 42
"How Google Took Over the Classroom" (Singer), 65–66
human rights, 13, 22, 34, 104, 112, 155, 160–61; algorithmic bias and, 51; Alston on, 50
Human Rights Commission, Australian, 51
Hungary, 173
hybrid weapons, 99–100
"La Hybris del Punto Cero" (Castro-Gómez), 38

Iberdrola, 155
ICC. *See* International Criminal Court
identity, 18, 44, 48, 128
IFTs. *See* Integrated Fixed Towers
immigration, immigrants and, 22, 31, 39, 99–101, 103–4, 115–16, 124
Immigration and Customs Enforcement, US, 103–4
imperialism, 24–25, 93–95, 123, 140–48, 188, 199; in Aotearoa-New Zealand, 168–69; digitized, 107–8; green, 157–62; marginalization and, 38
independent contractors, 126
India, 34, 38, 158
Indigenous and First Nations people, 11, 26, 167–68, 177–80, 185, 188, 202; dispossession, 42, 120, 138; education for, 68; industrialism impacting, 119; marginalization of, 38; rights of, 181–82; settler colonialism impacting, 141; sovereignty of, 23, 158, 166, 169–71

infinite growth, 148–53
Infinity Lithium, 135–37
information technology (IT), 71
infrastructure, 6–7, 17, 23, 55, 112, 186, 198; data analytics, 98–99; digital, 9, 63, 65, 73, 75; educational, 13, 28, 56, 64–66, 65, 72, 83; extractivist, 12, 142–43; imperialism and, 140–48; legal, 29, 42, 127, 161–62
Instagram, 18, 187–89
Integrated Fixed Towers (IFTs), 103
intellectual property rights, 188
International Criminal Court (ICC), 100, 175, 199–200
International Energy Agency, 176
International Labour Organization Convention 169, 181–82
International Monetary Fund, 153
internet access, 9, 56, 143, 145–46, 185
Interpol, 23–24
invisibility, 33, 43–45, 59, 142
iOS operating system, 185
iPhone, 144–45, 147, 186
Iran, 89
Iraq, 110–11
Ireland, 148, 192
Ironfield, Natalie, 165
Islamic State (ISIS), 89
Islamophobia, 47–50
Israel, 8–9, 19, 51, 85–87, 92, 99, 103, 113
IT. *See* information technology

Jackson, Moana, 165
Jacobin (journal), 131
James, C. L. R., 119
Jamil, Rabih, 128
Japan, 172
JEDI. *See* Joint Enterprise Defense Infrastructure
Jefferson, Brian, 97–98
Joint Enterprise Defense Infrastructure (JEDI), Google-Pentagon, 113
joint-stock companies, 24, 166

Kalpagam, Umamaheswaran, 38
Kerssens, Niels, 13
Khan, Lina, 193–94
Khashoggi, Jamal, 105
Kilgore, James, 98
Knox, Robert, 199
KPMG, 156–57
Kukutai, Tahu, 26
Kundnani, Arun, 41

labor rights, 117, 121, 125–26, 131
LAPD. *See* Los Angeles Police Department
Larkin, Brian Larkin, 140–41
Latin America, 38, 71, 125, 147, 152, 170–71, 202
Latinx people, 99, 115–16, 125
law enforcement, 13–14, 21–22, 97, 102–4, 111, 194
learning management system (LMS) technologies, 62
Lemkin, Rafael, 163
Leonardo (arms manufacturer), 155
LexisNexis, 104
von der Leyen, Ursula, 152, 158–59
LGBTQIA+ communities, 47, 188
liberal criticism, limitations of, 50–53
Lickanantay people, 179–81, 183
limited liability, 193, 201
liquidation, corporate, 200–201
lithium, 143, 157, 159, *159*, 161; mining, 12, 133–39, 147–48, 167, 176, 180–83, 203–4
Lithium Iberia, 133–35
Liu, Wenjuan, 180
Living Planet Index, WWF, 152
Lizardoia, Euskal Herria, 167
LMS. *See* learning management system
Loach, Ken, 130
lobbies, 112, 126, 155–56, 193–94
Local Call (Israeli media outlets), 85
Lockheed Martin, 88, 103
Loewenstein, Antony, 8
logistics centers, 7–8, 144

Lorey, Isabell, 118
Los Angeles Police Department (LAPD), 110–11
Luckacks, George, 199

machine learning, 38–39
Madrid, Spain, 2, 54, 71
Malm, Andreas, 10
Maloncy, Gavin, 130
management system (LMS) technologies, 62
marginalized communities, 21, 38–39, 49, 161, 185, 188
market dominance, 4–5, 7, 13, 125, 142, 192
market techno-solutionism, 152–57
Marx, Karl, 61, 120–23
Marxism, 19, 27–28, 96, 118–23, 198–99, 201
mass incarceration, 40–42, 46, 97–98
mass surveillance, 14, 192
materiality, 29, 42, 115–17, 124, 140–43, 186–87; González on, 96–97
mathematics, 38–39
Matviyenko, Svitlana, 96
McQuade, Brendan, 98–99
McQuillan, Dan, 14
means of production, 28, 115, 117, 121–24, 127, 205
Medina, Stuart, 156
Mediterranean Sea, 100
Mejias, Ulises, 15, 25–26
Melilla, Spain, 102
mental health, 18, 190–91
Messenger (Instagram and Facebook), 187
Meta, 7–9, 15, 29–30, 146, 186–92. *See also* Facebook
metadata, 105
Mexico, 88, 100–103, 170–71
Mezzadra, Sandro, 44
microchips, 143
microplastics, 172
Microsoft, 5, 63, 71, 113
migrant workers, 3, 120–21

Mijente collective, 101
Milei, Javier, 47, 138, *139*
militarization, 48, 102
military, 15–17, 99, 107–11; AI, 85–86, 89–95, 112; Israeli, 85–89; technologies, 19, 28, 110
minimum wage, 3, 125, 129
mining industries, 23, 155, 159–61, 170–73; deforestation and, 151–52; lithium, 12, 167, 176, 180–83, 203–4
mobile phones, 2, 5, 7, 21, 115, 185–86; extractivism and, 143–44
Mobile Video Surveillance System, Tactical Micro, 103
Molnar, Petra, 15, 26, 101
monetization, 6, 67, 122
monopolies, monopolistic behavior and, 16–17, 143, 177–78, 193–94
Morales, Evo, 138
Moreton-Robinson, Aileen, 42, 119, 141
Morrison, Scott, 37
Moten, Fred, 141
MuckRock, 13
multiculturalism, 41–42
Mumford, Lewis, 187
Muñiz, Ana, 26, 99
Musk, Elon, 9, 138, *139*
Muslim people, 47–50, 188
Myanmar, 188

Nakba of 1948, 87–88
national security, 16, 21, 87, 91–92, 97–99, 104, 111, 176
National Security Agency (NSA), US, 91, 105–6, 192
NATO. *See* North Atlantic Treaty Organization
naturalization of classroom digitalization, 75–79
natural resources, 23, 29, 174; extractivism of, 11–12, 133–40, 180–82; raw materials and, 12, 133, 139, 143–44, 158–60, 183
Nazi Germany, 22, 155

necropolitics, 15, 129–30
Neilson, Brett, 44
Neocleous, Mark, 99
neoconservatism, 31, 188
neoliberal, neoliberalism and, 3, 13–14, 27–28, 47, 169, 182; collapse of, 170; commodification, 66–67; criminology, 111; education impacted by, 67–71; extractivism and, 137; Global North and, 45–46; International Monetary Fund supporting, 153; police and, 97–98; precarity and, 118; racial, 27, 31, 35; racism, 40–43
Netanyahu, Benjamin, 9
Netherlands, 31–37, 41, 43, 45, 49, 52; Islamophobia in, 27, 41, 43, 48–49
Network Enforcement Act, Germany, 195
neutrality, 33, 73, 124, 128–29, 140, 148, 163, 178; absence of, 39–40
New Deal, US, 154–55
"New Jim Code," 43
New Latin American constitutionalism, 170–71
New Mexico, 81–82
New York University, 66–67
New Zealand, 34–35, 120, 167, 187
Next Generation Europe, 136
Ngāi Tūhoe people, 168–70
Nieves, Michelle Valentine, 115
Nimbus (Israeli cloud project), 113
Noble, Safiya Umoja, 21, 26
"No Child Left Behind Act," US, 70
North Atlantic Treaty Organization (NATO), 92–93
North Carolina, 143
No to the Cañaveral Mine Platform, 134
NSA. *See* National Security Agency
NSO Group Technologies (Israeli defense firm), 105

Obama, Barack, 91–92, 107
Obersavatorio de las Multinacionales de América Latina, 147

Offensive Cyber Operations, US, 107
O'Neil, Cathy, 39, 51
online education, 58, 61–62, 72–73
online platforms, 5–10, 194–97. *See also specific platforms*
open-source platforms and technology, 71, 132
operating systems, 185
oppression, 9, 22, 25, 35; algorithmic, 35–37, 45; Indigenous, 165–71; racialization and, 44, 119–20
Orban, Viktor, 47
Ovetz, Robert, 61–62

Pakistan, 10, 106
Palantir, 22, 51, 88, 104
Palestine, Palestinians and, 8–9, 51, 85–87, 99, 113
Panama, 144
Papua New Guinea-Bougainville, 155
Parliament, Dutch, 34, 43
Pashukanis, Evgeny, 20
Pasquale, Frank, 51
Pasquinelli, Matteo, 121
Pasternak, Shiri, 141
Patriot Act (2001), US, 91
Pearce, Frank, 19, 21
Pegasus spyware, 105
Pentagon, US, 96, 113
Personal Responsibility and Work Opportunity Reconciliation Act (PRWORA), US, 46
Phi4tech, 133, 139
philanthropy, 5, 56, 76, 177
piece wages, 127–28
Pinochet, Augusto, 66–67, 178
planned obsolescence, 24, 144–45
Planning tool for Resource Integration, Synchronization and Management (PRISM), NSA, 106, 192
Plataforma Salvemos la Montaña, 137
platform society, 5–10, 56–57
+972 (Israeli media outlets), 85

Poitras, Laura, 105
police, 25, 68, 95, 99, 104, 108–11, 169; automation and, 97–98; predictive policing and, 13–14, 21–22, 109–11. *See also* law enforcement
"Political Declaration on the Responsible Military Use of Artificial Intelligence," US, 94
political repression, 41–42, 48, 105–8, 113–14
politics, 187–88; algorithmic oppression and, 36–37; techno-solutionism and, 10–16
populism, 66–67
Porou, Ngati, 42–43
Porter, Amanda, 43, 165
Porter, Christian, 37
Portugal, 162
poverty, 13–14, 50, 108–10
power, 20, 37, 41, 42, 96, 126, 165; computational, 6, 15; corporate, 4, 16, 18, 55, 77, 104, 117, 200; infrastructural, 6–7; political, 20, 170, 188, 207; state, 31–34, 38, 198, 201, 204; of surveillance, 91
precarity, 1–4, 7–8, 100, 115–18, 120, 126, 130; of adjuncts, 59; neoliberalism and, 40; workers organizing against, 54–55
predictive policing, 13–14, 21–22, 92, 97, 109–10
predictive technologies, 36, 96–97, 108
PredPol (crime prediction tool), LAPD, 110–11
pregnancy, 130
premature deaths, 41, 44
preventive/data-driven systems, 36
PricewaterhouseCoope, 156–57
PRISM. *See* Planning tool for Resource Integration, Synchronization and Management
prisons, penal apparatuses and, 4, 40–41, 97–98
privacy, 80, 128, 189, 194, 205; violations, 6–7, 34, 81, 98, 192, 202

privatization, 4–5, 12–14, 28, 46–47, 82–84, 178; education and, 61, 63–83; healthcare, 69; infrastructure, 56; neoliberal, 67; protectionism and, 50
privilege, 6, 21–22, 96, 118, 120
Production Development Corporation (CORFO), Chile, 180
profits, 2, 4, 19, 54, 82, 190–91, 197; extractivism, 182
proprietary technologies, 5–6, 8, 56, 72–74, 127, 129, 195
"Protecting Kids Online" Senate hearing, 191
protests, 8, 54–55, 68, 113–14
Prussian Empire, 39
PRWORA. *See* Personal Responsibility and Work Opportunity Reconciliation Act
psychological experimentation, 189–90
psychological warfare, 97
public-private partnerships, 69, 80, 83
public services, 4, 12, 32, 46–47, 50, 71, 82–83
punishment, 4, 21, 31–33, 172
punitive apparatuses, 19, 26, 40, 98, 111, 196–98, 204; Maori population encountering, 42–43; neoliberalism and, 42, 47

quantification, 38, 60, 69–71, 78, 107, 109, 122
Quechua people, 170, 176
Quijano, Aníbal, 42

racial capitalism, 4, 39–42, 70, 97–98, 125–30
racialization, 3–4, 14–15, 21–22, 25, 119–21, 165; algorithmic, 31–32; colonial, 44, 112–13; machine learning and, 38–39; of Muslim people, 47–50; neoliberal, 27, 31, 35, 40–41, 50; settler colonialism and, 141

racism, 5, 21, 31–35, 109–11, 188, 205; algorithmic, 26–27; colonial, 38; colorblind, 44; Islamophobia and, 47–50; neoliberal, 40–43; state, 43–45
Ramirez, Mary Kreiner, 200
RAND corporation, 90, 110
rape culture, 24
raw materials, 12, 133, 139, 143–44, 158–60, 159, 162, 183
reactive/code-driven systems, 36
Reagan, Ronald, 46
Redón, Stella, 180–81
regulations, 8, 193–203
regulatory tolerance, 200
religion, 48
repression, political, 41–42, 48, 105–8, 113–14
repressive technologies, 95, 97–98, 111
responsibilities, 46, 70, 73, 126, 191, 197, 205; for algorithmic errors, 37; corporate, 40, 51, 83, 188–203; data management, 79; fossil capitalism and, 10–11; legal, 188–89; military, 89–95, 112
Richardson, Rashida, 26, 36
Riders x Derechos, 132
Rigakos, George, 99
rights, 19, 119, 188, 194–96, 202–3, 205; collective, 14, 17, 171, 197; of colonial subjects, 45; corporate, 24; digital, 80, 92, 94–95, 192, 195; to due process, 80–81, 106–7; Indigenous, 160–61, 181–82; "just war," 93; labor, 117, 121, 125–26, 131; of nature, 163, 166–72; of platform workers, 125–32. *See also* human rights
Rio Tinto mining, 150, 155
risk assessment, 21, 31–33, 101, 109–10, 159, 196
RMIT. *See* Royal Melbourne Institute of Technology
Robert, Stuart, 37
Robinson, Cedric, 42, 119–20
Robodebt, 37
ROBORDER initiative, European, 102

Rodríguez, David, 41–42, 165–66
Rodríguez Goyes, David, 165–66
Rohingya genocide, 188
Romania, 173
Rosenthal, Caitlin, 38
Royal Melbourne Institute of Technology (RMIT), 115–16
Rua, Fernando de la, 69
rural communities, 12, 14, 162–63, 166, 170, 202
Rutgers Institute for Information Policy and Law, 13
Rutte, Mark, 31, 52

Sadowski, Jathan, 26
El Salar de Atacama, Chile, 29, 176–83
salarization of production, salaries and, 121, 127
Salinas de Gortari, Carlos, 69
Samsung, 147
sanctions, 19, 175, 183, 192–93, 196, 200
San José Lithium min, 147–48
Saudi Arabia, 88
Save the Children, 86, 97
scandals, 15, 18, 64, 105–6, 188–203; SyRI, 31–34; welfare surveillance, 35
Schaeffer, Felicity Amaya, 15
Schmidt, Eric, 64
Schrems, Max, 192
Screen New Deal, 14
search engines, 7–8, 63, 194–95
securitization, 48, 201
self-governance, 169, 171, 181–82
Senegal, 185
Serbia, 155
settler colonial, 88, 99, 141
sexism, 188
Shilliam, Robert, 43
Sibelco, 143
Silicon, 143
Silicon Valley, 35, 44, 54, 56, 96
Singer, Natasha, 65
slavery, 24–25, 38, 119–20, 170–71

smart cities, 26
Snowden, Edward, 105–6
social fabric, 6, 55, 117, 180–81
social harms, 5, 13, 16, 19–21, 24
Socialist Workers Party, Spanish, 136
social networks, social media and, 5, 8, 87, 139, 187–203
Sociedad Química y Minera (SQM), 178–80
soft law, 173–74
soft power, 194
solidarity, 113, 132
Sorry We Missed You (film), 130
Soto, Nuria, 132
South America, 120, 170, 177–83
South Carolina, 146
southern criminology, 165–67
South Korea, 185
sovereignty, 96, 124; Indigenous, 23, 158, 166, 169–71; technological, 26, 194, 205
Soviet Union, 171
Spain, 76–77, 102, 123, 125–26, 132, 155; colonialism and, 120; lithium mining, 133–36, 147–48; Lizardoia, 167; national security commission, 16–18; Pegasus spyware cyberattacks, 205. *See also specific cities*
Spice, Anne, 139–41, 166
Spying on Students campaign, Electronic Frontier Foundation, 64
spyware, 105
SQM. *See* Sociedad Química y Minera
Srnicek, Nick, 26
standardized tests, 70
Starlink, 9
state, 1, 70, 95, 104; punitive power, 31–34, 38, 198, 201, 204; racism, 31–33, 43–45; sovereignty, 14, 92–93; terrorism, 8–9, 85; violence, 112, 188
statistics, 31–32, 38–39, 117
Stone, Christopher D., 163
Stony Brook University, 63–64, 65
Stop Ecocide organization, 164, 175

streaming services, 7
strikes, labor, 8, 131–32
structural racism, 27, 33, 52
Stuxnet virus, 89, 91
subaltern populations, 38–40, 42, 120–21, 126
Subcommittee on Consumer Protection, US Senate, 190–91
subjectivities, 118
suicide, 18, 31, 190
supply and demand, 127–29
supply chain, 23
surplus population, 125–26
surveillance, 4, 7–8, 17, 22, 26, 45–47, 97–98, 122; algorithmic, 86, 128; border, 101–4; capitalism, 56, 59; colonialism and, 88, 112–13; cyberwarfare and, 92, 105–8; incarceration and, 41; mass, 14, 117, 192; of Palestinians, 86–87, 99; of platform workers, 130–31; repression and, 41; of students, 77–78, 80–81, 83; welfare, 34–35, 45–47
sustainability, 5, 134–35 , 146–48, 153, 156–57
Sutherland, Edwin, 21
Systeem Risico Indicatie (SyRI), Dutch, 31–37, 43–44, 52
system failure, 32
systemic change, 201

Tactical Micro, 103
Tallinn Manual on the International Law Applicable to Cyberwarfare, NATO, 92–94
"The Taming of Chance" (Hacking), 39
Tapia, Luis, 171
Tauri, Juan, 42–43, 165–66
taxes, 8, 15–16, 31–34, 45, 52
taxi sector, 127–28
Taylorism, 25, 117
technologization, 15, 99, 101, 186
technology. *See specific topics*
techno-solutionism, 10–16, 52, 60, 152–57

telecommuting, 2, 12, 57
terrorism, 8–9, 21, 25, 85–86, 98–99, 106–8, 111, 169
Te Urewera, 167–70. *See also* Aotearoa-New Zealand
Texaco, 173
Thacker Pass, 167
Thatcher, Margaret, 66–67
Tianqi Lithium, 178
Tiqqun collective, 15, 89, 203
Tombs, Steve, 192–93, 199, 201
Torre Rangel, Jesús de la, 202
transnational corporations, 173
transparency, 50–51, 86, 94, 105, 133, 195
Treaty of Waitangi, New Zealand, 168
Tronti, Mario, 122
Trump, Donald, 91–92, 107
Tudge, Alan, 37
tuition, 68
"The Twin Transition" report (Ernst & Young, Microsoft), 156
Tzouvala, Ntina, 93

Uber, 125–26, 128, 146–47, 188
Uber Eats, 115
ubiquity, universality and, 4, 51, 185–87
Ukraine, 89–90
UN. *See* United Nations
UNDRIP. *See* United Nations Declaration on the Rights of Indigenous Peoples
unemployment, 3, 48, 126, 130, 133
UNEP. *See* United Nations Environment Programme
unions, labor, 3, 63, 115, 125, 130–31
United Kingdom, 10, 46, 92, 125–26, 130–32
United Nations (UN), 10–11, 13, 92, 150, 172–73
United Nations Climate Change Conference, 158
United Nations Declaration on the Rights of Indigenous Peoples (UNDRIP), 181–82

United Nations Environment Programme (UNEP), 172–74
United States (US), 15, 17, 51, 87–89, 105–6, 171–72, 192–94; automated decision-making systems, 13; communications towers, 145; facial recognition in, 21–22; GDP, 149; lithium mining and, 176, 178; Mexico and, 100–103; multiculturalism, 41–42; prison populations, 40–41
University of Auckland, 66
University of California, Los Angeles, 110
University of Chicago, 108–10
University of Melbourne, 68, 115
unlawful or wanton acts, 182–83
unpaid labor, 59, 119

value extraction, 45
Verizon, 105–6
very large online platforms, 194–97
Videla, Jorge Rafael, 66–67
Vietnam, 185
Vietnam War, 96, 163
Viljoen, Salomé, 26
violence, 14–15, 102, 109–10, 120, 126, 190; colonial, 87, 141; inequality and, 41; organized, 94; police, 68; state, 112, 188
Virden, Larry, 1
La Virgen de la Montaña, 136
visas, 101, 115, 120–21

Wacquant, Loïc, 40
wages, 57
wages, wage theft and, 8, 125
Wall Street Journal, 18, 189–91
Wang, Jacky, 120
war, 112–13. *See also* cyberwarfare
War Crimes of the Rome Statute, ICC, 100
warehouses, 25; Amazon, 1–2, 8, 55, 131–32
"war on terror," US, 47–48, 89, 99
Watego, Chelsea, 165
water, 11, 134, 145–46, 178–81
Watson, Irene, 141

weaponry, technological, 85–87, 95
Weis, Valeria Vegh, 165–66
welfare, 42–43, 48, 50, 68; surveillance, 34–35, 45–47
well-being, 13, 149
West Bank, occupied, 8–9, 86, 88, 103, 113
Western Europe, 24
WhatsApp, 187–88
"What to do with the Harmful Corporation?" (Tombs), 192–93
white-collar crime, 21, 50, 111
whiteness, 190
white possessive, 119
white supremacy, 41–42, 120, 165
Whittaker, Meredith, 26
Whyte, David, 143, 147, 199–201
Wilders, Geert, 48–49
Wolkmer, Carlos, 171

working class, 25, 62, 68, 113–14, 118, 121–22, 125, 140
World Bank, 147
World War II, 155
World Wildlife Fund (WWF), 152

xenophobia, 188
Xnet, 80–81
Xochipala (Mexico), 170–71

Yeung, Karen, 36
Yue, Neriah, 188

Zhengzhou, China, 144
Zuberi, Tukufu, 38
Zuboff, Shoshana, 26
Zuckerberg, Mark, 191
Zureik, Elia, 87, 99

ABOUT THE AUTHOR

AITOR JIMÉNEZ is Associate Professor at the University of the Basque Country and the International Institute for the Sociology of Law (IISL). He is Fellow of the Department of Criminology at the University of Melbourne and an Affiliate of the ARC Centre of Excellence for Automated Decision-Making and Society (ADM+S).

Printed in the United States
by Baker & Taylor Publisher Services